The Pacific Islands

NATIONS OF THE MODERN WORLD: ASIA

Published in cooperation
with the Asia Society

The Pacific Islands

Paths to the Present

Evelyn Colbert

WestviewPress

A Division of HarperCollins*Publishers*

Nations of the Modern World: Asia

Copyright © 1997 by Westview Press, A Division of HarperCollins Publishers, Inc.

Published in 1997 in the United States of America by Westview Press, 5500 Central Avenue, Boulder, Colorado 80301-2877, and in the United Kingdom by Westview Press, 12 Hid's Copse Road, Cumnor Hill, Oxford OX2 9JJ

A CIP catalog record for this book is available from the Library of Congress.
ISBN 0-8133-3286-9 (PB)

The paper used in this publication meets the requirements of the American National Standard for Permanence of Paper for Printed Library Materials Z39.48-1984.

10   9   8   7   6   5   4   3   2   1

Contents

Tables and Maps

Foreword

The plot of *The Pacific Islands: Paths to the Present* begins in the 16th-century at the point of contact with the West. Ever since such contact became routine in the 18th-century, many Westerners have equated these islands with Paradise. Author Evelyn Colbert widens our tunnel vision and corrects the old stereotypes with this erudite and informative work on the South Pacific. She delineates the geographic and cultural characteristics that distinguish its major island groups; traces the islands political transition from Western colonies to the mostly independent polities of today; describes issues of governance encountered first by colonial administrators and eventually by the islanders themselves; and discusses their dealings with the outside world, such as nuclear testing and the destructive exploitation of their natural resources.

The Asia Society owes special thanks to Dr. Colbert for her careful research, meticulous writing, and unlimited patience. Deborah Field Washburn and Karen S. Fein edited the volume, and Susan Sokolski worked diligently on the book's production. The Society would like to express its appreciation to Patricia Emerson for her careful and skillful copyediting and Patricia Loo for her attentive proofreading. Finally, thanks to Carol Jones, Lynn Arts, and their colleagues at Westview Press for their support of the project.

The views expressed in this publication are those of the author.

Nicholas Platt
President, Asia Society

Preface

This small book was conceived in ignorance—my own. In 1978 my responsibilities in the State Department's Bureau of East Asian and Pacific Affairs having been extended to the Pacific islands, I found myself entering a world for which three decades of largely East Asian experience had hardly prepared me. Nor was it easy to fill the gap from the writings of others. With the exception of a few travel books, most of the quite limited literature on the area had been written by specialists for specialists—largely of the anthropological persuasion.

Today much has changed. Historians, economists, and political and social scientists are joining anthropologists in providing what is becoming a diversified and substantial literature. Even so, it has seemed to me that a gap remains, that there would still be utility in a brief work introducing the general reader to the institutions, policy preoccupations, and international roles of the island polities as they have been shaped by their physical circumstances, their traditions, and their interaction with the West and Western institutions.

The debts I have incurred in the course of this enterprise are many and substantial. Not least are those I owe to colleagues who guided my initial Pacific islands education: Richard Holbrooke, who, as Assistant Secretary of State for East Asia and the Pacific, did much to activate U.S. policy in the South Pacific; the late John Dorrance, a pioneer among U.S. foreign service officers in the breadth and depth of his understanding of the region; and William Bodde, whose talents, energy, and empathy sped his remarkable transformation from Europeanist to man of the Pacific.

My great debt to scholars who have specialized in Pacific island studies is only partially revealed in my endnotes. I am deeply grateful also to the hospitality afforded me during visits to the region that enabled me to meet political and church leaders, educators, and a variety of other representatives of island society. Officials representing the island governments as well as Australia, France, Japan, New Zealand, and the United States have been a major source of information and insights. I am especially indebted to Professor Robert Kiste of the University of Hawaii, Robert Sutter of the Library of Congress, and U.S. ambassador to Papua New Guinea, Richard Teare, who read and commented most helpfully on the entire manuscript. I owe thanks also to Karen Scullen, who faithfully applied her superb word processing skills to a frequently illegible manuscript. Finally, I am grateful to the Asia Society for its support of this project and to its discerning and helpful editors, Deborah Field Washburn and Karen S. Fein. The errors are all my own.

Evelyn Colbert

The Pacific Islands

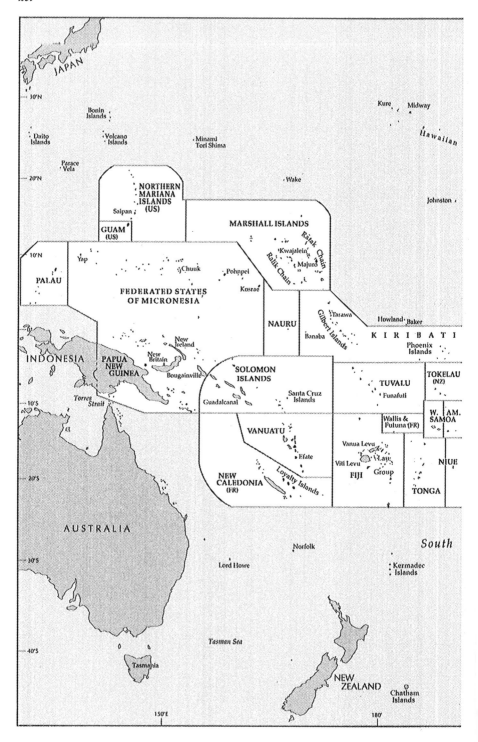

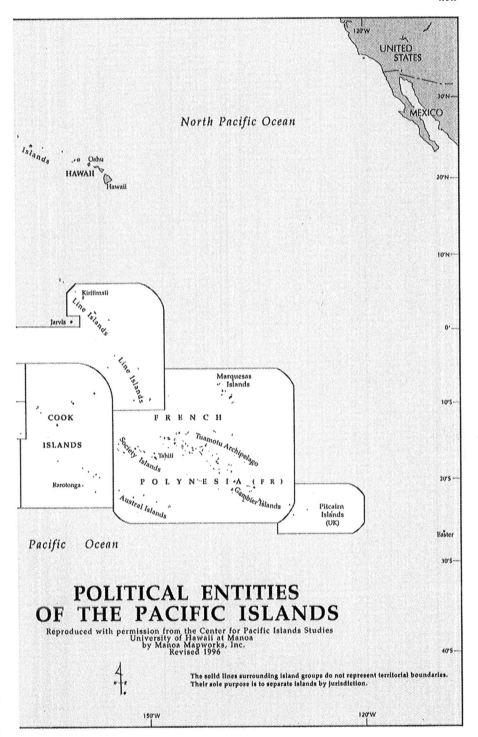

North Pacific Ocean

UNITED STATES

MEXICO

Islands

Oahu

HAWAII

Hawaii

Kiritimati

Line Islands

Jarvis

Line Islands

Marquesas Islands

COOK

F R E N C H

ISLANDS

Society Islands

Tahiti

Tuamotu Archipelago

Rarotonga

P O L Y N E S I A (F R)

Gambier Islands

Pitcairn Islands (UK)

Austral Islands

Easter

Pacific Ocean

POLITICAL ENTITIES
OF THE PACIFIC ISLANDS

Reproduced with permission from the Center for Pacific Islands Studies
University of Hawaii at Manoa
by Manoa Mapworks, Inc.
Revised 1996

The solid lines surrounding island groups do not represent territorial boundaries.
Their sole purpose is to separate islands by jurisdiction.

150°W   120°W

1

Introduction:
The Region and Its Peoples

Ten thousand islands—a few large, most very small, many uninhabited—are scattered over a vast oceanic expanse, stretching from the southern reaches of the Pacific to the Tropic of Cancer and covering 20 million square miles. The inhabitants of the islands, some 6 million people, see themselves in a unique relationship to their surrounding waters, one that distinguishes them in all their diversity as Pacific peoples.

Convention assigns the islands to a still larger region, Oceania, which includes Australia, New Zealand, and even Hawaii, whose claim to membership is based on its location, insular character, and the ethnic identity of its indigenous people. For the islands, however, Oceania is little more than a useful political device. It casts the much larger and more developed Australia and New Zealand as partners rather than patrons, while encouraging them to provide aid and protection. The inclusion of Hawaii reinforces island claims to U.S. attention.

Rhetorical concepts of an even more comprehensive Pacific grouping—whether the "Asia-Pacific Community," the "Pacific Basin," or the "Pacific Rim"—have had little resonance throughout the islands. Although anthropologists hold migration from Southeast Asia responsible for the original peopling of some of the islands, today's islanders, settled for millennia in their present homes, see little kinship between themselves and Asians, including those who have more recently taken up residence in their midst. Conversely, those outside the region who speak eloquently of a dawning Pacific century rarely have these small islands in mind.

In the everyday world, the islands command little attention. Deficient in most elements of global importance, whether economic, political, or military, they are also uniquely distant from the centers of international power and interest. In contrast, they have an unmatched aura in the world of the imagination. The Western image of Paradise on earth, originating from the accounts of eighteenth-century explorers, was perpetuated in the nineteenth century by writers and artists Paul Gauguin, Pierre Loti, Robert Louis Stevenson, and Herman Melville, among others. Today's tourist industry emphasizes the survival of this idyll, while journalists and social critics point accusingly to its demise under the impact of Western materialism.

Once under colonial rule, this thinly populated region is today the site of an extraordinary number and variety of mostly sovereign individual polities. Distance between island groups, different cultural characteristics and colonial histories, and the small scale and localism of precolonial societies have all been factors in creating this array. The 14 independent states form the largest group of political entities. They include Fiji, Kiribati, Nauru, Papua New Guinea, the Solomon Islands, Tonga, Tuvalu, Vanuatu, and Western Samoa, as well as five freely associated states—the Cook Islands and Niue with New Zealand and the Federated States of Micronesia, the Republic of the Marshall Islands, and Palau (Belau) with the United States.

The transformation from dependency to independence, in the Pacific islands as elsewhere, was a post-World War II phenomenon. However, decolonization began later than elsewhere and left more remnants of empire. These remants include the French territories of New Caledonia, French Polynesia, and Wallis and Futuna; the New Zealand territory of Tokelau; Chile's dependency, Easter Island; and, under the U.S. flag, the territories of Guam and American Samoa and the commonwealth of the Northern Marianas. Although these various dependencies are more fully self-governing than the typical colony of the age of imperialism, they lack essential elements of sovereignty and, in some cases, are still seeking changes in their status. Also remaining under Western sovereignty—U.S., Australian, or British—are islands that are minuscule in area and population even by South Pacific standards; for example, Britain's Pitcairn Island has a land area of 4.5 square kilometers and a population of 53, descendants of the *Bounty* mutineers.

Within the parameters of small size and dependency, the island polities vary widely in size, physical characteristics, and the number of islands that fall under their authority. The jurisdiction of some extends over hundreds of often widely dispersed islands; those of Kiribati, for example, are scattered over a distance equal to that between the eastern and western coasts of the United States. The jurisdiction of others extends over only one or two islands. In most island polities, whether compact or spread out, a substantial percentage of the population is clustered in and around a major urban center. Some island polities are the most mini of ministates. Only Papua New Guinea is large by global as well as local standards; occupying some

80 percent of the region's total land area, it is considerably larger than the United Kingdom or the Philippines. Papua New Guinea is the only country in the region that shares a land border with another state. As a legacy of the colonial era, the western half of the island of New Guinea, once part of the Netherlands East Indies, has been absorbed into Indonesia as the province of Irian Jaya.

Except in Fiji, Guam, and New Caledonia, the islands' inhabitants are mainly indigenous or migrants from elsewhere in the South Pacific. In Fiji, as of 1993, indigenes constituted only half of the population; for many years, they were outnumbered by Indians, mostly descendants of imported laborers, who now constitute 44.8 percent of the population. In Guam, only 42 percent of the population claims at least partial descent from the indigenous inhabitants, the Chamorro; Filipinos at 24 percent are the next largest group. In New Caledonia, the indigenous Kanaks make up only 43 percent of the population; the Caldoche, persons of European (mostly French) stock, constitute 38 percent; and Polynesians and Asians make up most of the remainder. In Fiji and French Polynesia, although citizens of European and mixed blood are not numerous, their communities are important: in Fiji because they mostly align themselves with the Fijians; in French Polynesia because of the prominence of the so-called *demis* in government and business.

Convention divides the indigenous peoples of the islands into three groupings: Polynesians, Melanesians, and Micronesians (see Figure 1.1). When explorers first used these terms they were primarily geographic categories: many islands, black islands, and tiny islands. However, they were soon applied to the different cultural characteristics of the peoples. As the study of island societies has progressed, many past generalizations and sharp distinctions have been challenged; in the words of the eminent Australian scholar Roger Keesing, these designations have come to have "somewhat ambiguous edges."[1] Nevertheless, they are designators with which the islanders have come to identify themselves and by which they are known to others, their oversimplifications and wholly European terminological origins notwithstanding.

Polynesia includes the Cook Islands, French Polynesia, Niue, the two Samoas, Tokelau, Tonga, and Tuvalu as well as Hawaii and New Zealand. There are also long-settled Polynesian pockets in parts of Melanesia. Polynesians are generally tall and of heavy build with skin color dominantly light or reddish brown. The languages of Polynesia are closely related; native speakers of one can quickly become fluent in another. Their traditional societies were highly structured along genealogical lines. The hierarchical order gave great authority to hereditary chiefs and attached to them some of the sacred power (*mana*) of the gods. But it also required some

4

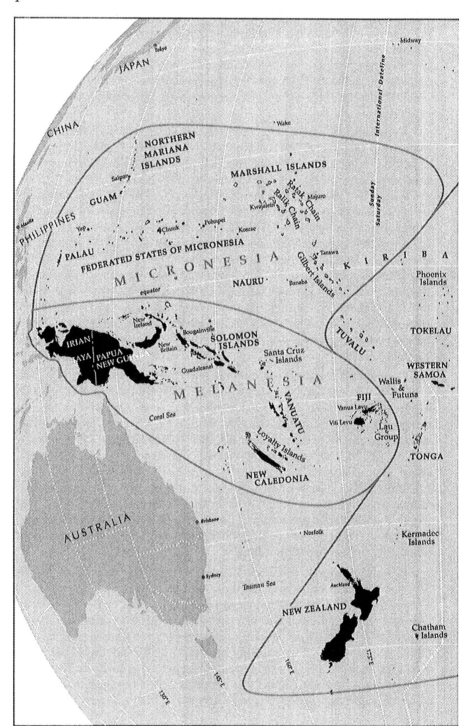

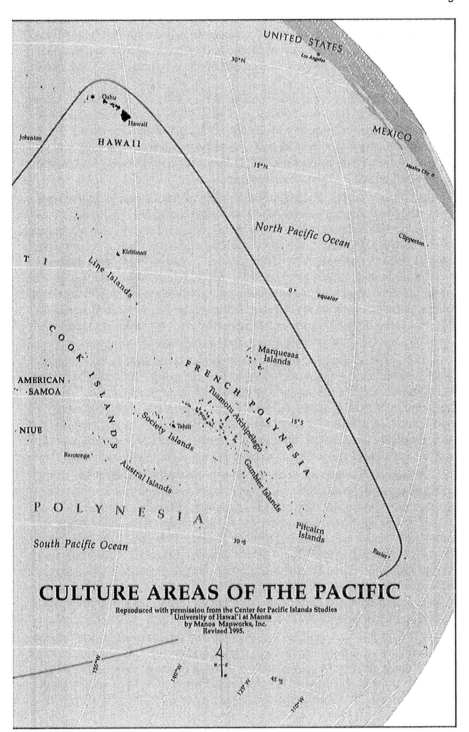

CULTURE AREAS OF THE PACIFIC

Reproduced with permission from the Center for Pacific Islands Studies
University of Hawai'i at Manoa
by Manoa Mapworks, Inc.
Revised 1995.

chiefly responsiveness to the views of those farther down the social ladder and to priestly interpretations of rituals and tabus. Rivalries among chiefs for status and power made for frequent warfare, usually brief and small-scale.

Melanesia includes the largest states of the region—Papua New Guinea and the Solomon Islands—as well as New Caledonia and Vanuatu. (Irian Jaya also remains dominantly Melanesian.) Fiji is Melanesian with a difference. Although geographically halfway between Melanesia and Polynesia, its chiefly society resembles that of its Polynesian neighbors, Tonga especially, with which it shares close historical links.

In contrast to the relative homogeneity of the far-flung Polynesians, great diversity characterizes the cultures of the less widely dispersed Melanesians. Scholars have attributed the development of shared characteristics in the Polynesian islands to the mobility of their seafaring inhabitants. Melanesians in contrast were disinclined to move outside their typically small-scale sociopolitical kinship units. Today Melanesia's peoples speak some 1,200 different languages, mostly mutually unintelligible; Melanesian pidgin provides something of a lingua franca for Papua New Guinea, the Solomon Islands, and Vanuatu. Melanesians also vary greatly in physical type and in skin color, which ranges from light brown to deepest black.

Traditional social organization also differed remarkably among the islands of Melanesia. Generally, however, societies were less stratified than in Polynesia, and lineage was less important in determining status. Thus the Melanesian Big Man achieved his position from his ability to amass a surplus of wealth he could share with his kinship group in the form of feasts and community buildings. Hostility to neighbors outside the kinship group was universal and the source of almost incessant small-scale warfare.

Micronesia is made up of the northernmost of the islands: Nauru, just south of the equator; Kiribati, north and south of the equator; and, north of the equator, Guam, the Northern Marianas, the Marshall Islands, the Federated States of Micronesia, and Palau. Micronesian characteristics are frequently described as falling between those of Polynesia and Melanesia. Like the Polynesians, the Micronesians were traditionally a seafaring people. In fact, they were the preeminent shipbuilders and most skilled and adventurous mariners of the South Pacific. Their traditional social structures, while differing significantly from one island group to the next, were generally hierarchical and based on lineage.

In the traditional societies of all three groupings, great significance was attached to propitiating a multitude of gods and spirits. Competitive giving and elaborate display were important elements in the pursuit of status, and the warrior ethic was held in high esteem. Title to land was vested not in individuals but in kinship groups whose members derived from their common ancestry a variety of obligations to one another. Economic life was village and family based; subsistence agriculture was supplemented by fishing, hunting, gathering, and some trade. With tools of bone, shell, and stone, islanders produced both the necessities of life and ceremonial objects, well suited to their purposes and pleasing to the eye. Into this traditional order, the West began to intrude on a significant scale only toward the end of the eighteenth century.

In the more than two centuries since these first contacts, the Pacific islands have undergone many changes. Some have resulted from the direct effects of the growing Western presence, bringing with it new commodities, forms of production, sources of authority, ideas, and religious beliefs. Others, more indirectly, have reflected changing events in the surrounding world. Empire building before World War II brought the islands into the global political order as dependencies of Western powers and Japan, a status islanders mostly accepted without significant resistance. Empire dissolution in the wake of World War II brought independence, or at least self-government, under political institutions derived from Western models but shaped by the continuing influence of tradition. Democratic institutions, introduced in the decolonization period, have been remarkably durable, despite a wide range of problems that mirror on a small scale those confronting other developing and even advanced countries. Whether sovereign or otherwise, most island polities remain economic dependents of those Western countries whose aid has been essential to compensate for meager resources and geographic isolation and to help meet the needs and expectations of growing populations. Their consciousness of weakness and dependence, however, has encouraged island governments to organize regionally, in order to influence policies pursued by the more powerful that affect island interests or well-being.

Note

1. R.J. May and Hank Nelson, *Melanesia: Beyond Diversity* (Canberra: Research School of Pacific Studies, Australian National University, 1982), Vol. 2, p. 2.

PART ONE

Exploration, Contact, and Control

2

Explorers, Traders, and Missionaries

The great age of European exploration in the fifteenth and sixteenth centuries had left the island world virtually unknown to the West and undisturbed by Western activity. By the time of the American Revolution, however, a new era of Western exploration had opened in the islands. Explorers, soon followed by traders and missionaries, brought the islands to the attention of a curious and fascinated West and introduced Western ways and goods to equally curious and fascinated islanders. The introduction of new products and technologies altered material culture; island cosmologies rapidly absorbed Christianity; and socioeconomic change and the involvement of Westerners impacted on traditional authority systems.

Explorers

The European explorers who followed Ferdinand Magellan across the Pacific were preoccupied with the search for routes to Asia. As their visits to the South Pacific were few and far between, they remained unaware of the multitude of inhabited islands scattered in its vast waters. Some sixteenth-century Spanish and Portuguese navigators landed in, or at least observed, one or another of the islands. But only in Micronesia was there a lengthy presence, in Guam, the southernmost island of the Marianas chain, which became a way station for the Spanish galleons that sailed from Mexico to Manila loaded with gold and silver for the China trade.

In some respects, these limited sixteenth-century contacts resembled those of later years. Encounters between Europeans and islanders were sometimes friendly but frequently hostile; whichever side initiated violence, Westerners quickly demonstrated the superiority of their firepower. Spaniards in pursuit of their own objectives took sides in tribal struggles.

Explorers claimed sovereignty for the Spanish Crown, and missionaries attempted to convert the islanders to Christianity.

Except in Guam and elsewhere in the Marianas, however, the Spanish period left little trace. For well over a hundred years the Western map of the South Pacific remained hopelessly inaccurate and largely empty, with the notable exception of Abel Tasman's discoveries in Australia and New Zealand in the mid-1600s.

All this was changed in little more than a decade—between 1767 and 1779—by the voyages of Captain Samuel Wallis, Admiral Louis-Antoine de Bougainville, and Captain James Cook. In the course of these voyages, Cook's three in particular, great strides were made in filling in the map of the Pacific islands and describing the islands and their peoples to the West. The notion of a Pacific Paradise became firmly embedded in the Western imagination as eighteenth-century explorers found in the islands, especially Tahiti, evidence of mankind's utopian existence in a state of nature.

Tahiti, the largest of the Society Islands, was far from the only island observed or explored. But it was the most completely examined over the longest period. The island held enormous attractions. With its high, green mountains, their slopes sparkling with waterfalls, its luxuriant vegetation and abundance of brilliant blooms, its freedom from dangerous animals and noxious insects, its even warmth moderated by sea breezes, Tahiti contrasted dazzlingly with the mariner's accustomed world—the long, dark, cold European winters, the hardship, dangers, and monotony of many months at sea. Its hospitable people were tall, well-formed, and graceful; extraordinarily clean by Western standards of the time, their skin the lighter tones of brown, they often matched and surpassed European standards of beauty.

In the West, the new discoveries aroused intense interest. The educated European audience had already been captivated by the theories of Jean- Jacques Rousseau ascribing superior virtues to man in a state of nature. The urge to expand the realm of knowledge and, above all, to organize it on rational principles was strong; the publication of Denis Diderot's 28-volume *Encyclopédie* began in 1751 and was completed in 1772, the year after the first publication of the *Encyclopedia Britannica*. At a less sophisticated level, a burgeoning popular press fed an insatiable public appetite for wonders.

The leaders of the expeditions into the South Pacific were truly men of the Enlightenment. Although commercial and political considerations were not wholly absent from their minds and their instructions, their dominant interests did not lie in trade or in altering the lives and beliefs of native peoples. They lay rather in adding to the world's stock of knowledge, comprehensively defined. Cook's first voyage was essentially a scientific expedition, sponsored by the Royal Academy as well as by the Admiralty. One of its principal purposes was to observe the transit of Venus across the sun (not expected to recur until 1874) in the mistaken belief that the data accumulated at Tahiti and other selected locations would make it possible to calculate the distance between the earth and the sun. Cook's companions included, as well as an astronomer, a number of naturalists and draftsmen. Also on board was the wealthy and well-connected young botanist Sir Joseph Banks (later the president of the Royal Academy for 40 years) with his entourage of naturalists and artists. Thousands of specimens and drawings came back to England with the explorers. The official account of Cook's first voyage, published in 1773 in three large illustrated volumes, quickly went into a second edition as well as an American one and circulated in French, German, and Italian versions. In the same years, Bougainville's *Voyage autour du monde* was also widely read.

While additions to knowledge were eagerly received, it was the explorers' romantic vision of island life that became most deeply embedded in the popular imagination. Bougainville thought that he had been transported to the Garden of Eden or, in classical terms, to a new Cytherea, where beautiful maidens emerged Aphrodite-like from the sea. The people, he wrote, "enjoy the blessings nature flowers liberally down upon them.... Everywhere we found hospitality, ease, innocent joy, and every appearance of happiness among them."[1] Banks found the scene "the truest picture of an Arcadia... that the imagination can form."[2] To the reports and drawings of the expeditions, poets and engravers at home added embellishments of their own. Thus Henry James Rye, future poet laureate of England, was inspired to write about the islands in 1783:

> Amid the wild expanse of southern seas
> Where the blest islands inhale the genial breeze
> The happier native in the fragrant grove
> Woos the soft powers of Indolence and Love[3]

Scientific knowledge of the island world continued to expand as explorers—American, British, French, and Russian—followed Cook to its waters and reported voluminously on their findings. But other visitors also followed. Some sought profit; others sought the salvation of souls; some achieved both. Western naval vessels became involved in these activities, sometimes punishing the islanders for crimes against their visitors, sometimes seeking to curb extremes of exploitative European behavior.

The Pacific islands were never a principal target of efforts to harness non-Western resources to the Western economy. Nevertheless, they offered a reasonable number of incentives to the venturesome trader and a whole new world of opportunity to the dedicated missionary.

Traders

In islands serving as ports of call, where gifts, trade goods, and services were exchanged, the material conditions of life were altered at an astonishing rate. In 1788 Captain William Bligh reported that, in the 11 years between his first and second visits to Tahiti, metal tools had completely replaced tools of stone and European pigs the native breed.[4] The growth of the whaling industry, bringing hundreds of vessels to island ports annually, increased the demand for sheltered anchorages, timber, fresh food and water, sexual favors, and locally recruited seamen. In 1844, at the industry's height, some 600 American whaling ships sailed Pacific waters.

Involvement in the international economy increased as traders found or introduced commodities that could be sold profitably elsewhere. Both collecting and producing were made more efficient when local trading companies became the intermediaries between island suppliers and itinerant trading vessels, replacing the earlier chiefs and beachcombers. As in galleon days, trade with China had repercussions in the islands. The European taste for tea and silk brought traders in search of sea otter pelts, sandalwood, and bêche-de-mer, these being among the few commodities other than specie and opium that the Chinese would accept in payment. In Australia a growing convict and free-settler population raised the demand for island foodstuffs, salt pork especially. As the demand for one product declined, that for another developed. The displacement of whale by coconut oil put an end to the islands' role as host to the whaling fleet. But they then became sources of coconut oil and centers of the local processing that produced the more easily shipped copra.

Much of the island world was unsuitable for plantation-type cultivation. However, temporary shortages of raw cotton during the American Civil War encouraged efforts to develop cotton plantations in Fiji. When the cotton boom ended and the plantations shifted to sugar, inadequate or unsuitable local labor supplies stimulated recruitment of contract labor, mostly from the New Hebrides (now Vanuatu) and the Solomons. Known as "blackbirding," the system gave rise to misrepresentation, brutality, and other abuses. But it also provided opportunities for young men to see something more of the world and acquire more of its goods, while fostering the growth of pidgin as the Melanesian common tongue. Blackbirding continued into the first decade of the twentieth century and, during its course, employed some 100,000 Melanesians in the islands and in Australia. The Melanesians who survived mostly went home; not so the Indian contract laborers who began to displace them in Fiji later in the nineteenth century.

Although involvement with the Western economy brought significant changes, the extent of change was limited by the minor opportunities most of the islands offered for Western profit or settlement. Only in New Zealand and Hawaii did European settlement and large-scale agriculture reduce the indigenous people to a very small minority and traditional land rights to insignificance. Elsewhere, the old persisted side by side with the new. Traditional patterns of dominance and deference remained important; life continued to revolve around family and village; subsistence economies remained the rule, even though augmented by cash cropping and paid labor. Remote and difficult areas—the interior of New Guinea for example—remained untouched. But for many islanders, lifestyles and work styles changed markedly.

The islanders bargained for desirable Western goods as eagerly as the traders did for marketable commodities. Metal and metal tools were in great demand and for good reason; according to one estimate, the replacement of stone by metal reduced by more than one-third the time required to provide the necessities of daily life.[5] Cloth, clothing, bottles, beads, guns, tobacco, alcohol, and new foods were also much desired. As these commodities became more accessible and widely distributed, old skills were abandoned and new dependencies were created. Contact with the West also resulted in an alarming decline in population. Peoples with strong warlike instincts could now kill one another with greater efficiency; tobacco and alcohol took their toll; and, most of all, imported diseases swept the islands in epidemic proportions.

Missionaries

The missionaries brought even more far-reaching changes. In Polynesia, conversion proceeded rapidly, and by the mid-nineteenth century Christianity was well established. The islands of Melanesia posed greater obstacles. It was more difficult to relate Christianity to indigenous belief systems; the climate was less hospitable and the hazards to life and health much greater; languages were more numerous and difficult; and Melanesian society lacked the degree of chiefly authority that, in Polynesia, facilitated the simultaneous conversion of whole communities. Even so, Christianity eventually prevailed throughout the islands. Indeed, in all of the Asia-Pacific area, only in the Philippines was Christianity to have the same universal and permanent impact.

The missionaries were dominantly evangelical Britons, inspired by the Wesleyan fervor that had first swept through England's lower and middle classes in the eighteenth century and that did so much to shape the moral tone of the Victorian era. Few missionaries were more dedicated and active than the representatives of the London Missionary Society, founded in 1795 with the express purpose of bringing the Word of God to the South Pacific.

Although the evangelicals dominated in numbers, the missionary impulse was not confined to the British dissenting churches. The Church of England also established a missionary presence; American Protestants of several denominations were particularly active in Hawaii and some of the Micronesian islands; German Lutherans and Roman Catholics were active in New Guinea; and, in a number of islands, French Roman Catholics competed with Anglo-Saxon Protestants for converts.

Unlike the traders, the missionaries went to the South Pacific intent on transforming it. Unlike the explorers, although viewing some characteristics of the islanders with a sympathetic eye, the missionaries were not inclined to accept island customs and behavior as demonstrating the virtues of natural man. To the contrary, as one of the most discerning among them, Thomas Williams, wrote of Fiji, "In these islands, the theory of those who teach the innate perfectibility of man ... has had a thorough test, resulting in most significant failure."[6] Others might see the extraordinary beauty and abundance of the islands as bound to affect favorably the character of the inhabitants. The missionaries did not. The Fijian, wrote Williams, "has hitherto seemed utterly unconscious of any inspiration of beauty, and his imagination has grovelled in the most vulgar earthiness."[7]

The association of sexual license with Paradise, distasteful to some even in the more tolerant eighteenth century, was totally repugnant to the missionaries of the nineteenth. They may have agreed with the conclusion of one of Bougainville's companions that islanders "recognize no other God than love." But they could not join in the Frenchman's admiring comment that they were "without vice, without prejudice, without needs, without strife."[8] On the contrary, in missionary eyes, the islanders were living in a grievous state of sin from which it was the Christian's duty to rescue them.

Sexual license and the worship of false gods were not the only fatal flaws the missionaries found in island life. Infanticide, tribal warfare, and cannibalism (practiced with particular enthusiasm in Fiji) were also targets of crusading zeal. Tales of cannibalism in missionary and other accounts particularly fascinated Western readers. They were further titillated by assurances that even more horrifying details had been withheld in deference to their delicate sensibilities. Thus images of cannibal islands joined the older and more lasting image of Paradise.[9]

Missionary tasks were all-absorbing and all-encompassing. Old evils must be eliminated, new virtues and values substituted. The naked must be clothed; promiscuous, provocative, and erotic behavior, ceremony, and images must be extirpated; cannibalism, slavery, infanticide, and tribal warfare must be stopped. Sabbath observance and regular church attendance must be enforced; the lessons of the Bible must be made familiar and compelling to all; and settlements must be rearranged, their inhabitants clustered around the church to encourage regular worship and, by example and authority, the maintenance of high moral standards.

Living close to their flocks, the missionaries translated the Bible into the local languages (for which they devised written forms), opened schools, produced textbooks, trained native teachers, catechists, and pastors, contributed to the introduction of new tools and techniques and, not incidentally, did their best to protect the islanders from loose-living and extortive traders and seamen.

* * *

As the Western presence grew, the island world, already changed by early contact with the explorers, was increasingly drawn into the much wider world of its visitors. Change was no longer being generated solely from within in slowly moving, strongly traditional societies. Instead, change was being speeded through interaction with representatives of distant societies, eagerly pursuing change themselves and achieving it at an unprecedented pace.

Notes

1. Carl Stroven and A. Grove Day, eds., *The Spell of the Pacific* (New York: Macmillan, 1949), p. 122.

2. Alan Moorehead, *The Fatal Impact* (New York: Harper and Row, 1987), p. 34.

3. Bernard Smith, *European Vision and the South Pacific* (Oxford: Oxford University Press, 1960), p. 90.

4. Ernest S. Dodge, *Islands and Empires* (Minneapolis: University of Minnesota Press, 1976), p. 105.

5. R.J. May and Hank Nelson, *Melanesia: Beyond Diversity* (Canberra: Research School of Pacific Studies, Australian National University, 1982), Vol. 2, p. 328.

6. Thomas Williams, *Fiji and the Fijians* (Suva: Fiji Museum, 1982), Vol. 1, p. 119.

7. Ibid., p. 113.

8. Moorehead, *Fatal Impact*, p. 46.

9. The cannibal island image has not entirely disappeared. Only a few years ago it turned up in the conservative French press when *Figaro*, echoed by others, described Kanak militants as "heating up their cooking pots." In quite a different vein, General Sitivini Rabuka, leader of the 1987 coup in Fiji, explained that Indians could never be more than guests in his country because God had given Fiji to the Fijians, while the missionaries had turned it "from cannibal land into Paradise." Ralph R. Premdas, "Fiji under a New Political Order," *Asian Survey*, Vol. 31, no.6 (June 1991), p. 543.

3

The Tides of Empire

Until quite late in the nineteenth century, Western empires showed little interest in formally acquiring sovereignty over islands so distant and so lacking in strategic importance. In the 1870s, however, a new colonial era began. The Western imperialist drive, already well under way elsewhere, began to extend to the South Pacific. In this more intense competition for sovereignty a united Germany—already strong commercially in the region—joined Britain, France, and the United States. Once the process began, it went forward rapidly. By the first decade of the twentieth century, all of the islands, some joined together and some divided by new boundaries, had come under one form or another of Western jurisdiction as colonies, protectorates, or territories (see Table 3.1).

The Evolution of Colonial Rule

Well before the 1870s, indeed from the earliest days of the Western presence, governments at home were pressed by their nationals in the Pacific to exercise jurisdiction in the islands. As a matter of course, explorers planted their national flags on islands they took credit for discovering. Traders appealed to home governments for protection against their foreign competitors. Missionaries sought home government action to protect native peoples against abuse by traders, blackbirders, and plantation owners. Protestant pastors (usually British) and Roman Catholic priests (usually French) looked hopefully to assertions of authority by their respective governments that might curb the incursions of the rival faith. Island leaders, too, challenged by domestic rivals or foreign pressures, appealed for protection to the more favored of the Western governments.

However, even while these demands were mostly rejected and the islands remained predominantly native ruled, Western governments sought to protect and control their nationals in the region. Naval vessels

TABLE 3.1
Chronology: The Expansion of Colonial Control in the South Pacific

1788	Britain establishes a penal colony in Australia; by the early 19th century, its claim to sovereignty over the entire continent is generally recognized by other Western countries.
1840	Britain claims sovereignty over New Zealand.
1842	France declares its sovereignty over the Marquesas chain and Wallis Island and establishes a protectorate over Tahiti.
1842	The Netherlands declares its sovereignty over Western New Guinea.
1853	France annexes New Caledonia.
1856	The Hamburg trading firm J.C. Goddefroy & Sohn establishes itself in Samoa, which becomes the center for the expanding commercial activities of subjects of the various German states.
1874	Fiji is ceded to Britain by its chiefs.
1880	France establishes its sovereignty over Tahiti and some of the other islands in the Society chain.
1884	Germany annexes the northeastern portion of New Guinea and the adjacent islands of New Britain and New Ireland.
1885	Germany declares a protectorate over the Marshall Islands.
1886	Britain and Germany divide eastern New Guinea and the Solomon Islands chain; Germany recognizes British claims to the Gilbert Islands (now Kiribati) and the Ellice Islands (now Tuvalu).
1888	Germany asserts sovereignty over Nauru.
1888	Britain establishes protectorates over the Cook Islands and Phoenix Islands.
1889	Spain assigns its claims to the Caroline and Marshall Islands to Germany.
1889	Britain establishes a protectorate over Tokelau.

1892   Britain establishes a protectorate over the Gilbert and Ellice Islands.

1893   American businessmen, supported by U.S. Marines, overthrow the Hawaiian monarchy and establish a republic.

1898   The United States acquires Guam, the southernmost of the Marianas, from Spain and annexes Hawaii.

1898   Western New Guinea is incorporated into the Netherlands East Indies.

1899   Samoa is divided between the United States and Germany (U.S. possession of American Samoa is formalized in treaties with local chiefs in 1900 and 1904).

1901   Britain establishes a protectorate over Tonga and transfers Niue to New Zealand.

1905   Britain's sector of New Guinea is transferred to Australia as the Territory of Papua.

1906   Britain and France establish joint rule (the condominium) over the New Hebrides (now Vanuatu).

1919–20 The post–World War I settlement places former German possessions under the League of Nations mandate system: in Micronesia, to be administered by Japan; in New Guinea by Australia; in Nauru to be under the joint control of Britain, Australia, and New Zealand.

were important symbols of interest, presence, and power. Their commanders performed extensive public functions, among them mediating disputes and making semidiplomatic agreements. Thus Charles Wilkes, leader of the U.S. Exploring Mission of 1838 to 1842, negotiated commercial agreements with the principal chiefs of Samoa and Fiji, setting out the terms on which islanders and traders were to deal with one another. Naval commanders also imposed punishments for crimes against their fellow citizens, punishments that could include even execution and the destruction of entire villages.

A more continuous presence was provided by foreign consuls, whose involvement in tribal power struggles and competition for influence was frequently among the sources of local disorder. Another continuous presence was provided by the missionaries, whose influence in local affairs extended well into the political realm. As advisers to island chiefs they produced law codes and constitutions combining Western concepts of order and equal justice with requirements for church attendance and sabbath observance.

Britain, with its unmatched command of the seas, its growing settlements in Australia and New Zealand, and its forward shipping base in Sidney's magnificent harbor, was the dominant Western presence in the Pacific from the earliest days of contact. But in the nineteenth century France, not Britain, took the first steps to bring Pacific islands under colonial rule. Far behind Britain in their long-standing rivalry—driven out of North America, with only a tiny foothold in India, forestalled in hopes of establishing itself in New Zealand—France was more interested in acquiring island possessions in the Pacific than either Britain or the United States. In 1842 Louis Philippe, inspired by the desire to revive some of the glory of the Napoleonic era and urged on by French missionaries, annexed the Marquesas and established a protectorate over Tahiti, both, in due course, to be incorporated in French Polynesia. Ten years later France annexed New Caledonia.

It was not until 1874 that Britain followed in French footsteps and annexed Fiji. Before then, it had remained satisfied with a policy that ensured equal access for its traders and missionaries, maintained the jurisdiction of the Crown over British subjects, and, when feasible, responded to the pressures of the moral activists for curbs on slave trading and other abuses. Adding the administration of these remote islands to the burdens of empire in India and elsewhere recommended itself to neither the Treasury nor the Colonial Office. The Australian and New Zealand colonies, much closer to the scene, might argue a strategic as well as commercial and missionary interest. But their reluctance to share the expense involved in bringing Pacific islands into the empire reduced still further their ability to inspire the British government to action.

Cook, in his day, had been as skeptical of the value of island acquisition as the British government was to be in the century following his voyages. Concerning the potential value of Tahiti, he wrote, "Notwithstanding nature hath been so very bountiful to it yet it doth not produce one thing of intrinsic value or that can be converted into an Article of trade, so that the value of the discovery consists wholly in the refreshments it will always afford to shipping in their passage through the seas."[1] Almost a century later, the British government responded similarly to pressures for resisting French claims in New Caledonia, seeing no grounds for the argument that the islands were "so important to Her Majesty's Colonies and to British Commerce in these Seas that their occupation by the French cannot be allowed."[2]

In 1874, however, Britain changed its course. In that year Prime Minister Benjamin Disraeli accepted the pleas of Fijian chiefs for annexation. Only a year earlier, his predecessor, William Gladstone, had refused in traditional terms "to be a party to any arrangement for adding Fiji and all that live beyond it to the cares of this overdone and overburdened Government and Empire."[3]

Britain's acquisition of Fiji was in some sense an ad hoc response to the Fijian problem—the evident need for some accepted authority to curb the encroachments of British settlers and to moderate the conflict among the chiefs that the ambitions of Cakobau, the most powerful among them, had done much to intensify. But in abandoning well-established precedents, Disraeli, an avowed imperialist, was also reflecting the heightened drive toward enlarging overseas empires, already evident in Africa and Asia.

Joining the competition, worldwide and in the Pacific, Germany, united in 1871, sought to catch up with the other great powers and assert its public authority in support of what was already an extensive private commercial presence. The United States also, with the Civil War behind it, was in a better position to assert its interests and support those of its nationals. These had become particularly strong in Hawaii where, in 1893, members of the U.S. business community overthrew the Hawaiian monarchy and established a republic. Although U.S. officials, consular and naval, had been implicated in these events, President Grover Cleveland resisted appeals for annexation. By 1898, however, U.S. reluctance to acquire overseas colonies had been overcome sufficiently to make Hawaiian annexation acceptable at home. In that same year, the United States not only added Guam to its Pacific island holdings but also acquired Puerto Rico and the Philippines. The division of Samoa between the United States and Germany followed not long after.

Even as the islands were being divided among the Western powers, however, the urge to acquire new possessions and deny positions of advantage to rivals or potential adversaries was balanced by the desire to avoid confrontation in low-priority areas. Compromises that balanced the entitlements of Western powers were often achieved by dividing or exchanging islands or island groups. Such arrangements did not often reflect the best interests of the islanders, who were rarely consulted and had little or no influence over the outcome. At times agreements to exchange or divide island territories were less reflections of the local scene than by-products of agreements centered on other more inflammable parts of the world. Thus, in the late 1880s, in the wake of agreements moderating the high tensions of Anglo-German competition in Africa, new arrangements were made in the Pacific. Sovereignty over the western half of the island of New Guinea, long in Dutch hands, remained undisturbed; the eastern half was divided between Germany and Britain, Germany receiving the northern sector and the outlying Bismarck archipelago, Britain the southern sector, which, in 1906, it ceded to the newly established Australian federation as the territory of Papua.

The Samoan islands, inhabited by people much more homogeneous than those of New Guinea, were also partitioned in the interest of resolving Western differences. By the 1880s Samoa had achieved a position of some importance as the hub of the extensive German commercial interest in the islands and as the scene of a good deal of British and U.S. business activity. The Western business community was centered in the town of Apia, which, as a Pacific island port, had become second in importance only to Honolulu.

Outside Apia, however, the complex rules of the Samoan sociopolitical system, the Fa'a Samoa, remained untouched. Based in the extended family (*aiga*) headed by family chiefs (*matai*), its elaborate patterns of descent, decorum, deference, and decision making applied to most of life's activity. Fa'a Samoa supported a strong sense of common identity. But although chiefly titles were distinguished by gradations of importance, there was no single apex of authority, and high chiefs competed strenuously for status and power. Chiefly competition proved disastrous in the face of large-scale European land acquisition for plantation purposes, as it both reduced the ability of the aroused Samoans to deal effectively with the Europeans and also contributed to the disorder that threatened Western business interests and intensified their pressures for government intervention. The initiatives and intrigues of the U.S., British, and German consuls did nothing to moderate local disorders. Meanwhile, tensions among the rival Western states were being heightened by competitive treaty making and displays of naval power.

In 1889 Britain, Germany, and the United States agreed on elaborate arrangements intended to preserve Samoa's independence in principle, while giving their consular representatives authority over matters of Western interest. However, although the chiefs assented to the new arrangements, neither peace nor stability resulted. In 1899 new treaties were negotiated in which Samoa was divided between Germany and the United States. In return for Britain's renunciation of all its claims to Samoa, Germany renounced its claims in Tonga and Niue in favor of Britain, while also ceding to Britain the Solomon Islands east and southeast of Bougainville.

In 1906 still another compromise moderated European conflict in the New Hebrides while subjecting the islanders to a unique and unwieldy division of power. The constant conflict in the New Hebrides between British missionaries and traders and their French counterparts was too remote from the interests of London and Paris to be allowed to cast a shadow over the recently established Entente Cordiale, which had resolved more hotly contested colonial differences between the two in the Middle East, Africa, and Asia. Accordingly, in this small island group, an Anglo-French condominium established two separate colonial administrations, exercising authority not only over their respective nationals but over the Melanesian inhabitants as well.

On the eve of World War I, the division of the South Pacific islands among the Western powers was complete. Britain, France, Germany, and the United States were the South Pacific colonial powers as the twentieth century opened; Australia and New Zealand were soon to inherit some of Britain's colonial responsibilities. After its defeat by the United States in the Spanish-American War, Spain had given up all its claims in the region, not only ceding Guam to the United States but also ceding its claims in the Caroline and Marshall Islands to Germany in exchange for financial compensation. In Hawaii and New Zealand the indigenous Polynesians were now small minorities, in Hawaii already outnumbered by migrant Asians; in both Europeans dominated politics and economics. New Zealand's elevation to dominion status in 1907 and Hawaii's first step toward statehood when it became a U.S. territory in 1900 constituted the final act in their decisive differentiation from the rest of the South Pacific. In Melanesia, Western New Guinea's status as part of the Netherlands East Indies was unchallenged, although the Dutch were barely a presence there.

Of the other islands, only Tonga retained a form of sovereignty. In 1893 its ruler, King George Tupoua, passed down to his son, George II, the kingdom he had united under a constitution drafted by the Wesleyan missionary Shirley Baker. George II was plagued by financial difficulties and worried about the possible effects on Tonga of growing international rivalries. In 1900 his request for British protection was granted in a treaty making Britain responsible for Tonga's defense and foreign affairs but leaving Tongan domestic authority largely untouched.

World War I was a distant event to Pacific island peoples, with only a very few, a Fijian labor battalion among them, having firsthand contact with the hostilities. Nevertheless, Germany's defeat altered patterns of subordination to outside powers once again. Early in the war, Australian and New Zealand forces, meeting no resistance, took over German possessions in Samoa, Nauru, and New Guinea; the Japanese did the same in Micronesia. Post-hostilities arrangements legitimized these transfers not as colonies, but as mandates held under League of Nations authority.

Patterns of Colonial Administration

In most of the Pacific island dependencies, the pre-World War II colonial powers did not go very far in assuming broad functions and instituting bureaucratic machinery. Nor were Western-style representative institutions introduced. The Europeans exercised complete authority at the top over matters that concerned them; at the local level much of the customary approach to decision making was left undisturbed. In general the European presence tended to grow in proportion to plantation agriculture under settler control. Outside areas settled by Europeans, especially where access was difficult or conditions unhealthy, as in New Guinea and the Solomons, administrative machinery was rudimentary. Almost everywhere education, health, and welfare remained largely the province of the churches.

The specific policies of the administering powers varied with their objectives, the characteristics of the colony, and often, the particular character of the man in charge on the ground. Generally the British were the most concerned to protect native interests against settler and other encroachments and to maintain chiefly systems. This was especially the case in Fiji, the most important of their Pacific island colonies. There, the chiefly system (dominated by the eastern chiefs, among whom Cakobau was preeminent) was maintained and indeed strengthened by the organization of a Council of Chiefs, which in time became the official voice of the Fijian people and the apex of a hierarchical, village-based "native administration." At the same time, however, like other administering powers, the British strove to eliminate some of the practices, such as slave holding and intertribal warfare, on which chiefly power rested.

France was intent in New Caledonia on introducing European settlers and advancing their interests at the expense of the indigenous Kanaks. In the Polynesian possessions, largely unsuited to commercial agriculture and hence unattractive to settlers, the French administration was largely a tax-collecting machine. German practice varied similarly. In New Guinea, where a plantation economy of some consequence had developed in the coastal areas, the protection of settler interests dominated. In Western Samoa, to the gratification of the first governor, Dr. Wilhelm Solf, there was little expectation of significant contributions to imperial coffers. Accordingly, wrote Solf, his congenial duty was "merely to guard it as what it is—a little Paradise—and to do my best to keep the passing serpent out of our Garden of Eden."[4] Under both Solf and his successor the government discouraged foreign settlement and restricted land sales.

The United States, formally committed to preserving Samoan law and custom and protecting traditional land rights, left the administration of American Samoa and Guam to the navy, which had its principal impact on the islanders in programs for combating illiteracy and improving public health. In Samoa traditional leaders followed traditional patterns in performing such political and administrative functions as were of no concern to the naval governor and his staff. Beginning in 1905 the *matai* selected 25 of their number as members of an advisory council, or Fono. In Guam an appointive advisory council was established in 1917; it became elective in 1930. When, however, the Guam Congress petitioned for U.S. citizenship, the secretary of the navy testified to the U.S. Congress that the Guamanians had "not yet reached a state of development commensurate with the personal independence, obligations, and responsibilities of United States citizenship."[5]

In their dealings with the islanders they ruled between the wars, the Japanese were neither as dismissive as the French nor as protective as the British. "Our duty," said the first Japanese governor-general, "is to promote the welfare of the inhabitants of these islands. While an alteration of their manners and customs is not our sole objective, conservation of their old customs and folkways is not important either. It is our hope that the racial and ethnic differences will vanish so that the inhabitants can completely become our brethren at the earliest possible date."[6] Turning Micronesians into brethren meant educating them in the Japanese spirit. That they were also provided with electricity, sewage systems, new roads, and harbors was incidental to the endeavor to maximize the islands' contribution to Japan's economy by making available to large numbers of Japanese settlers (eventually to outnumber the Micronesians) the land and facilities needed for productive agriculture and efficient marketing.

Wherever resources were available to support plantation agriculture or mining, colonial administrations were faced with conflicts between such responsibility as they accepted for protecting indigenous peoples and the pressures of their nationals for access to land, labor, and profits. In dealing with this problem, Britain, in Fiji, chose what seemed to be a singularly enlightened course; in New Caledonia, French policy was nothing if not exploitative. Paradoxically, however, the policies of both had negative long-term consequences persisting to the present time.

In Fiji British administrators as well as Fijians interpreted the commitment Britain had made to the interests of the Fijian chiefs and their peoples as a commitment to the paramountcy of Fijian interests over European. Accordingly, the British administration discouraged European settlement, prohibited alienation of land, established the jurisdiction of a Land Claims Commission over such land transfers as had already taken place, and stipulated that leaseholds could run for only 21 years.

However, in seeking to insulate the Fijians from settler exploitation, the British administration endowed them with a long-term problem. The need for plantation labor was great, and, if Fijians were precluded from this role by their preferences and by protective British law, another source had to be found. Thus began in 1879 the subsidized migration of Indian workers, many of whom preferred remaining in Fiji to returning home. The permanent Indian community, supplemented by free migrants, grew accordingly. By the turn of the century Indians in Fiji outnumbered Europeans by about ten to one. The doctrine of Fijian supremacy as a shield against European encroachments had become a shield against the possibility of Indian predominance. In 1904, when the purely advisory Legislative Council was enlarged to include non-Europeans, two seats were accorded to Fijians and none to Indians, who did not achieve any representation until 1916.

British policies were based on the determination that Fiji should not become a settler colony. French policies were based on the expectation that New Caledonia, in the Australian-New Zealand pattern, could become a prosperous overseas home for their citizens. Grazing, plantation agriculture, and later nickel mining seemed to provide opportunities for development sufficiently attractive to lure French settlers, whose supply of native labor would be augmented by transported convicts.

In this vision land expropriation was the key to development. The Melanesian inhabitants, the Kanaks, then believed to be a dying race, had neither a place in the vision nor any rights. As the minister of the navy and colonies wrote in 1854:

> The uncivilized inhabitants of a country have over that country only a limited right of domination, a sort of right of occupation.... A civilized power on establishing a colony in such a country acquires a decisive power over the soil, or, in other terms, she acquires the right to extinguish the primitive title.[7]

The "right" was freely exercised, as was its indispensable component, the "right" to confine the Kanaks to reservations, or *cantonnements*, where they were required to remain by the *indigénat*, the regulations governing the indigenous people.

As a settler colony, New Caledonia never lived up to expectations. Its resources proved insufficient to attract settlers numerous enough to ensure that the Kanaks, like the Maori and the Hawaiians, would be reduced to a very small minority. Even so, the lives of the Kanaks were transformed.

By the opening of the twentieth century, the Kanak *cantonnements* occupied only 16 percent of the land of New Caledonia's largest island, the Grande Terre, mostly of the poorest quality. Kanaks were still in the majority, although not by much, at 53 percent of the population versus the 40 percent constituted by the Caldoche. But the Kanaks lost even this majority when the nickel boom and French encouragement produced a sizable flow of migrant workers, mostly Polynesians. For the Kanaks, the land was elemental—"the blood of the dead." Losing it meant losing identity, losing contact with ancestors, losing the very basis of family and tribal organization.

Having lost their birthright, the Kanaks gained no corresponding benefits from their dependent status. The fruits of development were enjoyed largely by the Caldoche concentrated in Noumea and its environs. French authorities rarely concerned themselves with the well-being of the natives for whom only the missionaries provided schools and health care.

From the earliest days of colonial rule sporadic Kanak violence against French settlers and officials reflected the anger aroused by land alienation, forced relocation, labor and other abuses, and the depredations of foraging cattle. A wave of such violence culminated in 1878 in a major uprising led by High Chief Atai. In the course of the rebellion, suppressed with great brutality, 1,200 rebels were killed and large numbers of villages were razed. Thereafter rigorous French controls effectively curbed remaining resistance.

The 1878 uprising, along with another, smaller-scaled rebellion in 1917, became important events in the Kanak nationalist tradition. Chief Atai has become a legendary figure as has Louise Michel, the French Communard transported to the New Caledonia penal colony who has been credited with teaching the rebels how to cut telegraph wires.

The Kanak uprisings were a major exception to the general absence of significant organized protest against colonial rule in the pre-World War II Pacific islands. The Mau movement in Western Samoa was another. The two differed significantly. The Kanaks were fighting for the past against the already advanced destruction of the Melanesian traditional order. In Western Samoa, where much of the traditional system had been incorporated in the colonial administration but with limits on its autonomy and authority, resistance to some forms of change was mixed with demands for fuller participation on Samoan terms in the institutions introduced by the Europeans. In New Caledonia, tribal rivalries weakened resistance to the French, and Kanak chiefs fought in alliance with the French as well as against them. In Western Samoa resistance to colonial authority was much more widely supported and coherently organized, reflecting both the well-established sense of Samoan identity and the mobilization capabilities of Samoa's pervasive, village-rooted authority system.

Pressures originating under German rule for a larger Samoan voice continued when jurisdiction was transferred to New Zealand after World War I. New Zealand administrators, mainly drawn from military ranks, ignored Fa'a Samoa and local policies and ruled with a heavy hand. In 1927 discontent took organized form in the O le Mau (Samoan League). The Mau declaration, in familiar Western liberal terms, asserted the equality of "all men ... in the sight of God" and the ruler's obligation "to assist the members of a subject race in advancement towards ... a government of the people in accordance with the will of the people."[8]

Employing nonviolent and passive resistance tactics, the Mau quickly won the support of both traditional leaders and villagers and were able to organize a police force and collect taxes. To the New Zealand administration, however, Mau tactics—refusal to pay taxes, demonstrations, uniforms, badges, and boycotts—seemed to foreshadow more threatening actions, and it was as determined as the French to resist challenges to its authority.

It ordered the Mau to disband, arrested and exiled Mau leaders, prohibited uniforms, and restricted or banned demonstrations. Police units often quite brutally sought to root out the movement and its supporters; in one encounter in 1929, 11 Samoans were killed, among them holders of several of the highest titles. The Mau were equally stubborn, and the stalemate was broken only in 1935 with the electoral victory of the New Zealand Labor Party and the first steps toward fulfillment of Labor's pledge to promote Western Samoan self-government.

New Zealand's commitment to change, however, had no parallel elsewhere in the islands, where the status quo remained almost unchallenged. On the eve of Japan's attack on Pearl Harbor in December 1941, the South Pacific remained a placid backwater in the Western view. Two hundred years of contact with the West had brought many changes to island life. But these changes, like the foreign presence itself, had been unevenly distributed and had had irregular effects on traditional patterns. For the European or American measuring changes in the islands against those that had transformed the West in the same two centuries, it was easy to conclude that Paradise, for better or for worse, had remained well rooted in its traditional ways.

Notes

1. W.P. Morrell, *Britain in the Pacific Islands* (Oxford: Oxford University Press, 1960), p. 16.

2. Ibid., pp. 103–104.

3. Ibid., p. 161.

4. Michael J. Field, *Mau: Samoa's Struggle for Freedom* (Auckland: Polynesian Press, 1991), p. 26.

5. Ruth G. Van Cleve, *The Office of Territorial Affairs* (New York: Praeger Publishers, 1974), p. 85.

6. Ron Crocombe and Ahmed Ali, eds., *Foreign Forces in Pacific Politics* (Fiji: University of the South Pacific, 1983), Vol. 4, p. 116.

7. John Connell, *New Caledonia or Kanaky* (Canberra: Australian National University, 1987), pp. 42, 43.

8. J.W. Davidson, *Samoa Mo Samoa* (Melbourne: Oxford University Press, 1967), p. 119.

PART TWO

Toward Independence
and Elsewhere

4

The Catalyst of War

The Second World War opened a new era in the South Pacific. Prolonged and costly battles brought the region to world attention as never before; places hitherto hardly known—Guadalcanal, Tarawa, Rabaul, the Kokoda Trail—figured in daily headlines. A large Allied presence, in which Americans were by far the most numerous, imposed greater pressures on traditional attitudes and ways of life. Such Allied pronouncements as the Atlantic Charter aroused expectations of a new postwar order of freedom and justice. When the war ended for the rest of the world, the islands returned to the obscurity from which they had so briefly emerged. But the war's impact on the islands helped to shape their future.

Hostilities in the islands were as fierce as in any World War II theater; in February 1944 the most intense artillery barrage of either world conflict preceded the Allied landing on Kwajalein, Japan's military headquarters in the Marshalls. Little of the region was untouched, whether by bombing, ground conflict, or logistical use. Almost simultaneously with their attack on Pearl Harbor, the Japanese had moved from their military bases in Micronesia to capture Guam from the United States. By the middle of 1942 they had made landings in force at various points in New Guinea, the Gilberts, and the Solomons. However, Japanese plans to carry their campaign to the islands farther south and to Australia had been disrupted by crippling air and naval losses in the Battle of the Coral Sea in May and at Midway in June. Thereafter the sea and air counteroffensive was joined by large-scale Allied landings in 1943 in New Guinea, the Gilberts, and the Solomons. Meanwhile, the air strikes on the supply facilities and strong points of both sides brought death and destruction to islands untouched by ground warfare.

Involved in the war against Japan from the outset, the Southwest Pacific was engaged until its very end. Even in 1945, after the Japanese had been forced out of most of the islands, ground hostilities continued in New

Guinea where more than 100,000 Japanese remained when the war ended. Other Pacific islands also were engaged until the war's last days. The Micronesian island of Tinian, taken by the United States in July 1944, served as launch site for the nuclear attacks on Hiroshima and Nagasaki (as it had previously for almost 30,000 B29 sorties) while some 20,000 U.S. servicemen were gathered there to participate, if need be, in a ground assault on Japan.

Japanese occupation, Allied liberation, and all that lay between brought great devastation and suffering. Labor was conscripted on a large scale in islands occupied by the Japanese, and in some cases workers were forcibly transported to other islands where the need for their labor was greater. Continuing hostilities brought grave shortages to Japanese-occupied islands; the consequent struggle for food turned the Japanese from the stern taskmasters of prewar Micronesia to brutal exploiters. Japan's occupation and its subsequent expulsion left some areas in ruins; by the time the United States recovered Guam, its economy had been completely destroyed.

Islanders were not merely victims of the hostilities; many became active participants. Although most served with Allied forces, some from Micronesia fought under the Japanese flag. Fijians enlisted in British forces in large numbers; the Fijian First Commandos fighting in the Solomons were noted for their bravery, ferocity, and skill in jungle warfare. Islanders from New Caledonia and French possessions in Polynesia, serving with the Free French Pacific Battalion, fought in North Africa, Sicily, and southern France. Cook Islanders joined New Zealand forces. The Australian-officered Pacific Island Regiment, in which more than 3,500 Melanesians had served by the war's end, fought in most of the major New Guinea campaigns. Australia's Coast Watching Organization relied heavily on islander assistance in its surveillance of Japanese movements and rescue of downed Allied airmen. Wartime military service, recognized by citations and decorations, contributed both to islander self-esteem and European respect for islander prowess.

In islands largely untouched by fighting on the ground, the war in many cases brought new prosperity, employment opportunities, and improved infrastructure. At the same time, however, populations were uprooted to make way for military facilities; Allied demands for labor—whether volunteer or conscript—denuded villages of able-bodied men; and bombing attacks on military strong points drove islanders from their villages and gardens into the bush.

The Allies constructed huge bases in the New Hebrides, New Caledonia, Fiji, and Samoa; lesser installations, like those memorialized in James Michener's *South Pacific*, were widely scattered. In some cases new airfields, roads, or harbor facilities gradually disappeared with the end of their military use. In others they survived to supplement the rudimentary prewar infrastructure. Thus, Honiara became the capital of the Solomon Islands to take advantage of the proximity of Henderson Airfield and other installations left by the military.

Prewar islander contacts with Europeans, in a system of strict social segregation, had been largely confined to colonial officials, missionaries, and local entrepreneurs. Now in many cases islanders had seen their European overlords put to flight by the advancing Japanese and encountered enormous numbers of Allied troops. By mid-1942 U.S. marines constituted almost 15 percent of the population of Western Samoa's principal island. By 1943 more than half a million Allied troops were serving in Melanesia as compared with a European presence of 31,000 in 1941. Over one million U.S. servicemen passed through the island of Manus in the Admiralties, the largest Allied base in the South Pacific. These contacts had substantial impact. The lure of Western goods and the cash economy was reinforced by what seemed to be the unlimited wealth of the Allies and the generosity with which Allied servicemen distributed it. The presence of black troops had its own impact; segregation in the U.S. armed forces was of less consequence to islanders than their perceptions that black men like themselves were on a par with white men in their uniforms, arms, and possession of worldly goods.

Islanders were struck by the egalitarianism and generosity of U.S. troops. Transients without the vested interest in maintaining the acceptance of their superiority that was such a strong pillar of the colonial system, U.S. servicemen largely ignored orders against fraternization and giving gifts. The manner of their giving was important also, fitting well into the island ways of cementing relationships. Testifying to the novelty of American behavior, a Solomon Islander said, "They invited us inside [their tents], and when we were inside we could sit on their beds. We got inside and they gave us their glasses so we could drink out of them too. They gave us plates and we ate with their spoons. This was the first we had seen of that kind of thing."[1]

Inevitably, out of these new experiences ideas developed about future relations between islanders and Europeans. In most cases these new thoughts entered the atmosphere only somewhat amorphously. But in a few they found more organized expression. In the Solomons, after its recapture in 1943, the Maasina or Marching Rule movement organized paramilitary groups whose mission was not only to reconstruct the villages but also to resist the restoration of British rule. In postwar Tahiti the new spirit prompted the organization of trade unions and of a committee seeking local autonomy. In the Cook Islands newly organized trade unions evolved into a movement calling for autonomy.

Ideas were changing in the West as well. Elsewhere in the colonial world during the interwar years the issues of self-government and independence had been debated to some effect; representative institutions had been established in some British colonies, while the Philippines was moving toward promised independence. But the idea that the tiny islands of the South Pacific might someday govern themselves was hardly entertained. Indeed the islands that had become League of Nations mandates after World War I were assigned to the class of dependencies whose minuscule size, small population, and remote location made them seem likely to remain permanently integral parts of the territory of the administering power.

However, with the region so much in the wartime public eye, with Australia and New Zealand energetically insisting upon a voice in any Pacific settlement, and with the idea of liberation very much in the air, the future of the islands demanded attention at least as a subset of the broader colonial issue. President Franklin D. Roosevelt's sweeping but rather vague pronouncements suggested that, if Washington had its way, the unhampered authority of colonial powers would be replaced everywhere by an international trusteeship arrangement, applicable to Allied as well as enemy colonies, to ensure that colonial peoples were being prepared for self-government and for independence if that was what they wanted.

Although Roosevelt's view of trusteeship paralleled that of Labor-led governments in Australia and New Zealand, it was hardly in accord with the views of British Prime Minister Winston Churchill or Free French leader, General Charles de Gaulle. Churchill heatedly dismissed the notion that Britain's colonies could fare better under an international system and declared, "I will have no suggestions that the British Empire is to be put into the dock and examined by everybody to see whether it was up to their standards."[2] For General de Gaulle restoring the French Empire was a first and essential step in restoring the prestige France had lost at Vichy.

By the time decisions had to be made, U.S. policy had also moved away from the idea of an all-encompassing international supervisory system. For the defense establishment and its supporters on Capitol Hill, wartime experience had demonstrated the key role of the islands—those of Micronesia in particular—in the defense of the Pacific, a task now falling largely to the United States. The casualties suffered in the campaign to expel the Japanese—over 6,000 American dead, almost 25,000 wounded—intensified military emotions. While the military campaign in Micronesia was still under way the chief of staff, Admiral William Leahy, argued, "The conquest of the islands is being effected by the forces of the United States and there appears to be no valid reason why their future status should be the subject of discussion with any other nation."[3]

To those in the State Department and elsewhere who argued for consistency with the U.S. anti-colonial tradition, Secretary of War Henry L.

Stimson responded that acquisition of the Japanese mandates would "not represent an attempt at colonialization or exploitation ... it is merely the acquisition by the United States for the defense of the Pacific for the future world. To serve such a purpose they must belong to the United States with absolute power to rule and fortify them."[4]

Senator Mike Mansfield, among other congressmen, advanced much the same argument.

> We need these islands for our future defense, and they should be fortified whenever we deem it necessary. We have no concealed motives because we want these islands for one purpose only and that is national security.... No other nation has any kind of claim to the Mandates. No other nation has paid the price we have.[5]

The domestic argument having been settled in favor of the military position, the United States could not easily press its allies to submit to the international controls over their still extensive possessions that it was unwilling to accept for itself. Accordingly, the UN trusteeship system was applied only to former mandates. In the South Pacific these included New Guinea under Australian trusteeship, Nauru under Britain, and Western Samoa under New Zealand, all responsible to the General Assembly and the Trusteeship Council for the implementation of trusteeship obligations to foster welfare and progress toward self-government among the indigenous peoples. The former Japanese mandate, now under U.S. authority as the Trust Territory of the Pacific Islands (TTPI), was uniquely designated a strategic trust. Under this arrangement, the United States was free to establish military bases on the islands, station its troops there, and exclude others. Oversight was entrusted to the Security Council rather than the veto-free General Assembly.

Although most dependencies in the Pacific and elsewhere remained outside the formal trusteeship system, the principles of trusteeship were given wider application in Article 73 of the UN Charter, the Declaration Regarding Non-Self-Governing Territories. In all dependencies, UN members, accepting their obligations as a "sacred trust," were not only to promote the economic, social, and educational advancement of their wards but also "to develop self-government, to take due account of the political aspirations of the peoples, and to assist them in the progressive development of their free political institutions." The obligation was largely a moral one. The sensitivities of the colonial powers were reflected in the limits placed on their required annual reports to the secretary general, which were to be confined to economic, social, and educational conditions and to be "subject to such limitations as security and constitutional considerations may require." Even with these limitations, however, the charter provided a new basis for bringing pressures to bear to speed the decolonization process.

Also included in Article 73 was the injunction to colonial powers to cooperate with one another in their efforts to advance social, economic, and scientific development in their dependencies. In 1947 the South Pacific Commission (SPC) was organized for these purposes at the initiative of Australia and New Zealand. Although its original membership was limited to the colonial powers and its functions to research, advice, and assistance, the SPC came to provide a valuable training ground in regional cooperation for island leaders, whose voice in the organization became increasingly important as progress toward self-government proceeded.

Notes

1. Geoffrey M. White and Lamont Lindstrom, *The Pacific Theatre: Island Representations of World War II* (Honolulu: University of Hawaii Press, 1989), p. 10.

2. James F. Byrne, *Speaking Frankly* (New York: Harper and Bros., 1947), frontispiece.

3. William D. Leahy, *I Was There* (New York: Whittlesey House, 1950), p. 210.

4. Evelyn Colbert, *Southeast Asia in International Politics* (Ithaca: Cornell University Press, 1977), p. 48.

5. Philip W. Manhard, *The United States and Micronesia: A Chance to Do Better?* (Washington, D.C.: National Defense University, 1979), p. 25.

5

Patterns of Political Evolution

In the South Pacific, as elsewhere, World War II had shaken the foundations of empire. However, the transition from colonial subordination to self-government and independence came relatively late and unevenly to the islands, leaving some still in dependent status but with high levels of autonomy (see Table 5.1). The decolonization process, although slow, was comparatively peaceful, consensual, and smooth. But it was not exclusively so. Separatism was a serious problem in several of the emerging states. In Micronesia arriving at mutually satisfactory relations between islanders and Washington, although a peaceful process, was an unusually prolonged and contentious one. Only New Caledonia suffered from the violent clashes between colonial authorities and militant nationalist movements so common in other parts of the world.

The British Connection: Empire Dissolved

By the war's end Britain, under Labor rule like Australia and New Zealand, had accepted independence as the goal of its colonial policy. However progress toward that goal was slow. Colonial administrators tended to advise a long period of tutelage. Island leaders might favor a faster pace, but they too accepted the need for gaining experience in self-government and for time to frame constitutional arrangements that would be widely understood and supported. Western Samoa in 1962 was the first to emerge as an independent state. The process peaked in the 1970s. However, it did not end for Britain until 1980 when the Anglo-French condominium was terminated and the New Hebrides entered independence under the name of Vanuatu.

By 1980 the Commonwealth decolonization process had resulted in nine independent states and two states freely associated with New Zealand. Only New Zealand's Tokelau remained a dependency. Retaining old ties,

TABLE 5.1 The Polities of the South Pacific

	Political Status	Capital	Principal Islands or Island Groups
American Samoa	U.S. territory	Pago Pago	Tutuila
Cook Islands	independent in free association with New Zealand	Avarua	Rarotonga
Federated States of Micronesia*	independent in free association with the United States	Kolonia	Yap, Chu'uk, Pohnpei, Kosrae
Fiji*	independent	Suva	Vita Levu, Vanua Levu
French Polynesia	French territory	Papeete	Society Islands, Tuamota Archipelago
Guam	U.S. territory	Agana	Guam
Kiribati	independent	Tarawa	Gilbert, Line, Phoenix Islands
Marshall Islands	independent in free association with the United States	Majuro	Enewetak, Kwajalein
Nauru	independent	no official capital	Nauru
New Caledonia	French territory	Noumea	Grande Terre
Niue	independent	Alofi	Niue
Northern Marianas	U.S. territory	Saipan	Tinian

Palau*	independent in free association with the United States	Koror	Babelthuap
Papua New Guinea*	independent	Port Moresby	New Guinea, New Britain, Bougainville
Solomon Islands*	independent	Honiara	Guadalcanal, Choiseul, Gizo, Santa Isabel
Tokelau	New Zealand territory	none	Nakunonu, Atafu, Fakaofo
Tonga	independent	Nuku'alofa	Tongatapu, Ha'apai, Vaváu
Tuvalu	independent	Funafuti	Funafuti
Yanuatu*	independent	Port Vila	Espiritu Santo, Malakula, Efate
Wallis and Futuna	French territory	Mata-Utu	Ile Uvea, Ile Futuna
Western Samoa*	independent	Apia	Savai'i, Upola

Member of the UN.

the new states became members of the Commonwealth and continued to receive economic assistance from their former rulers.[1]

The institutional steps along the path to independence differed very little from one Commonwealth dependency to another. Step by step the once purely advisory Legislative Councils, dominated by ex-officio and European members, were enlarged to increase their indigenous membership. Election by universal suffrage replaced appointment, membership became dominantly or wholly indigenous, and, finally, a cabinet system was established along Westminster lines, with the head of state a primarily ceremonial figure, whether appointed by the Crown or locally elected.

As the process moved toward completion, indigenous representatives joined Commonwealth officials in establishing a constitutional basis for the new state, and by various means the views of the population were consulted before the formal proclamation of independence. In the Solomon Islands, for example, over a two-year period a draft constitution was submitted for comment first to local assemblies and then to the popularly elected National Assembly, while more than 100 meetings acquainted the people of the various islands with the constitution's terms. An equally lengthy process and extensive public participation preceded the adoption of Papua New Guinea's constitution. In Western Samoa, the people were not directly consulted during the drafting, but the proposed constitution was submitted to a UN-observed popular referendum in accord with trusteeship requirements. In Fiji the British, Fijian, and Indian drafters met behind closed doors in order to shield efforts to accommodate conflicting ethnic interests from popular passions.

Although representative bodies elected by universal suffrage were at the heart of most of these constitutions, in some cases special arrangements protected traditional systems. Tonga was the most extreme case. Restored to full sovereignty in 1970, it retained the royal authority of the 1875 constitution; commoners elected only a minority of a legislative assembly dominated by the royalty and nobility. The constitution of Western Samoa confined the right to vote and to be elected to *matai*.

In Fiji elaborate constitutional provisions protected Fijian political and economic rights at the expense of Indians, the numerical majority. Despite pressures for an equal franchise, the framers adopted a complex communal voting system. Candidates ran on one of three lists—Fijian, Indian, or General Elector (citizens of European, part-European, and Chinese descent). Each elector cast four votes, two for candidates of his own community, one each for candidates of the other two. In the lower house 22 seats were reserved for Fijians, an equal number for Indians, and 8 for General Electors.

The system did not absolutely rule out an Indian-dominated government. However, elaborate safeguards, including an appointed upper house with a virtually automatic Fijian majority, made it next to impossible to amend the constitution or alter long-standing arrangements for preserving Fijian land tenure and local autonomy structures. The Great Council of Chiefs, still the instrument of the chiefly hierarchy although enlarged to include high-ranking public officials and other notables, not only appointed a third of the senators but was also empowered to veto any legislation affecting Fijian rights and customs.

Elsewhere although equal suffrage was the rule, traditional values were also recognized. Chiefs were often given some role, even if a purely formal one. Customary land rights received particular attention, reflecting both their defining role in the island sociopolitical order and fear of the temptations presented by the opposing Western concept that treated land as a commodity like any other.

Much earlier, Tonga's 1862 law code had epitomized the depth of island concern with land alienation, decreeing:

> It shall in no wise be lawful for a chief or people in this Kingdom of Tonga to sell a piece of land to foreign people... it is verily, verily forbidden for ever and ever; and should anyone break this law he shall work as a convict all the days of his life until he die, and his progeny shall be expelled from the land.[2]

In the same spirit, if not with the same rigor, contemporary island constitutions protect traditional title to land in one form or another and severely restrict its alienation whether for public or private use.

Not infrequently Christianity, despite its more recent introduction, was among the traditions singled out for protection. The Tuvalu constitution, for example, defines the foundations of the state as "Christian principles, the Rule of Law, and Tuvaluan custom and tradition."[3] In similar terms, the constitution of Western Samoa declares that the state is "based on Christian principles and Samoan custom and tradition."[4]

Secessionism was a problem in a number of cases. In groupings that foreign rule had joined together, the process of constructing new nations was bound to raise questions of identity. Thus as the colonial period drew to a close, ethnic diversity, local loyalties, and concern for the loss of local resources asserted themselves in separatist movements.

For the Gilbert and Ellice Islands, which had become a single British protectorate in 1892, separation was indeed the simple answer. The not quite 50,000 Micronesians of the Gilberts had no desire to share their phosphate resources with the Polynesians of Ellice or to permit their continued predominance in the public service. For their part the 8,000 Polynesians of the Ellice Islands feared subordination to the six times as many

Gilbertese. Accordingly, the Ellice Islands became independent as Tuvalu in 1978, the Gilberts as Kiribati a year later.

In other places, there were no such easy answers. The thousand or so former inhabitants of war-devastated Banaba, now resettled in Fiji, opposed the island's inclusion in Kiribati as another of the indignities imposed by colonial rule and by the exploitation of their island's rich phosphate resources to the benefit of British companies and the revenues of the Gilbert and Ellice colony. Although in the end most of the Banabans remained in Fiji, an ingenious compromise gave them some privileges in the Kiribati constitution, which restored Banaban ownership of land taken over for phosphate mining, guaranteed the right of the Banabans to return to their home island, and gave Kiribati citizenship and representation in the House of Assembly to those remaining in Fiji.

In the Solomon Islands secessionist pressures also caused trouble. Secessionist sentiments were strongest in the western islands, somewhat more prosperous than their neighbors and anxious to preserve their resources for themselves. Elsewhere in the Solomons demands peaked on the eve of independence for a more equal relationship between center and provinces. The issue remained unresolved in 1978 when the Solomons became independent. A division of powers reasonably satisfactory to all concerned was not achieved until 1980.

The most difficult problems arose in Papua New Guinea. Their diverse colonial experiences had done nothing to prepare for single nationhood peoples with strong particularist traditions, separated by formidable mountain barriers and long distances at sea, and speaking close to a thousand mutually unintelligible languages. While members of the emerging elite shared educational experiences and the Christian faith, Michael Somare, the "father of PNG independence," was one of the few who put the idea of a united country ahead of provincial loyalties.

The closer independence came, the more local attachments became the basis of political organization. Advocates of strong provincial powers or even independence were especially active in Papua and in easternmost New Britain, where they cited cultural and behavioral differences as reasons for rejecting union with New Guinea. Nowhere, however, were secessionist pressures stronger than on the island of Bougainville, more closely linked geographically and ethnically to the Solomons than to the PNG mainland.

As PNG independence neared, Bougainville's resistance to incorporation in the new state increased. Appealing to the visiting Australian prime minister for a referendum, separatists disclaimed indigenous ties to Papua New Guinea. "To us they are aliens. We have nothing in common with them."[5] Economic factors also fanned separatism. The initiation in the mid-1960s of commercial exploitation of what promised to be huge copper deposits had both vastly magnified Bougainville's economic potential and convinced Bougainvilleans that wealth rightfully theirs was being diverted to foreign companies and the government in Port Moresby. Increasingly also Bougainville was suffering from the ills resulting from mining operations—environmental degradation, infringements on traditional land rights, and the presence of large numbers of unruly migrant laborers from the mainland. Almost simultaneously with PNG's independence day in September 1975, Bougainville seceded as the North Solomons Republic. Violence was avoided and unity preserved, largely owing to the efforts of Prime Minister Somare and Bougainville leader Father John Momis, whose yearlong negotiations resulted in an arrangement for strengthening provincial authority that became a model for the other provinces. In the semifederal structure that emerged, each of Papua New Guinea's 18 other provinces, like North Solomons Province, had its own elected assembly and cabinet and considerable authority over provincial affairs.

The system was not destined to survive. It seemed to have done more to provide employment—19 premiers, over 130 cabinet members, and some 500 assembly members—than to satisfy either opponents or proponents of decentralization.

Defenders of provincial autonomy attributed their difficulties to central government interference and indifference to local concerns. In Bougainville, once again the bellwether, armed secessionism erupted in 1988 and, as assaults on the provincial system intensified, the leaders of the four other island provinces threatened to join the secessionists in Bougainville in an independent state. Despite these threats, empty so far, and sharp divisions within the national cabinet, in June 1995 an overwhelming parliamentary majority approved the outlines of a new provincial system to be fully in place by the end of 1997. The new system both strengthens the authority of the national government and alters the provincial government structure, substituting for the former elected assemblies made up of the heads of local governments and the provincial members of parliament, one of whom will serve as governor.

In the New Hebrides Condominium, widely known as the "pandemonium," secessionism and other indigenous fissures combined with British and French differences to make the path to independence disturbed and painful. Britain, which ultimately prevailed, wanted to leave, but under the condominium agreement could do so only if the French left also. France wanted to stay or at least leave behind in support of Francophone interests a weak central government and strong provincial ones. In local politics Anglophones and Francophones eyed one another with deep suspicion, looking to their respective patrons for support as they jockeyed for preeminence. Religious differences also entered into political choices, Roman

Catholics, mostly Francophone, contending with the mostly Anglophone Protestant majority.

The strongest of the political parties emerging in the early 1970s was the Anglophone Vanuaaku Pati (VP) led by Father Walter Lini, an Anglican priest. In close alliance with the dominant Presbyterian church, the VP built a uniquely sturdy and pervasive local structure as the foundation for an efficiently organized national leadership. The Francophone parties were weak by comparison. Accordingly they sought alliances with two so-called custom movements—one on the island of Tannu, the John Frum cargo cult, the other, Nagramial on Espiritu Santo—both attached to the preservation of tradition and both resentful of Presbyterian power.

In 1978 Lini, dismayed by the lack of progress toward independence, established a provisional government that, although without legal status, brought independence negotiations to a conclusion, drafted a constitution, and in 1979 organized elections in which the VP won 62 percent of the popular vote. As in Papua New Guinea, separatists greeted the approach of independence scheduled for July 1980 with a secession attempt. In May, supported by French interests and the U.S. libertarian Phoenix Foundation (not incidentally also engaged in the real estate business), Nagramial forces declared the independence of Espiritu Santo, seizing its principal town and airport. Large-scale looting and intimidation followed, forcing thousands to flee. Not until the end of August were troops volunteered by Papua New Guinea and logistically supported by Australia able to bring the situation under control.

The French Connection: France *Outre Mer* Preserved

Not untouched by the reforming spirit of the war years, French leaders also felt obliged to show appreciation for colonial participation in Free French forces. Accordingly under the 1946 constitution indigenous inhabitants of French overseas territories, formerly French subjects, became citizens entitled to vote for the president of the republic and members of the Senate and the Chamber of Deputies in Paris, as well as for members of local assemblies. In New Caledonia the *indigénat* was abolished.

But independence was no part of postwar plans. Frustrated at great cost in efforts to retain its colonial empire in Asia and Africa, France met with greater success in the South Pacific. There, although forced to relinquish its share of authority in the New Hebrides, it succeeded in retaining its other holdings. However, the relationship has been transformed in many ways as France has responded, albeit often slowly and grudgingly, to demands for greater autonomy. Citing their territorial autonomy, universal suffrage, and the same full range of civil liberties and equal justice enjoyed by French citizens everywhere, the French argue that the remaining overseas territories can no longer be described as colonies. They are parts of France that happen to be overseas.

In two of the three French Pacific territories this argument carries considerable weight. In French Polynesia the combination of expanding autonomy and economic dependence on France has confined support for independence to a small minority. In tiny and deeply conservative Wallis and Futuna it does not exist at all. In New Caledonia, on the other hand, it is only since the late 1980s that the confrontation between Kanak *indépendentistes* and Caldoche loyalists has softened and conservatives in Paris have joined their socialist political rivals in recognizing concessions to the Kanaks as a means of preserving the French tie.

French Polynesia

The absence of communal tensions has eased French Polynesia's progress to its present levels of autonomy. A climate inhospitable to European agriculture made it easy for the French to fulfill their commitment to respect existing land titles. Persons of wholly European blood constitute only about 11 percent of the population. Somewhat more numerous and extremely influential in politics, business, and administration are the so-called *demis*, whose Polynesian blood is mixed with Chinese or European. The territory's two most recent presidents are *demis*—the indestructible Gaullist Gaston Flosse and the technocrat Alexandre Léontieff. Although generally more assimilated to French culture than the *maohi*, who count themselves as pure Polynesian, the *demis* have increasingly come to stress their Polynesian identity.

The postwar political scene, activated by expanded local authority and suffrage, was dominated by a strong proponent of local autonomy, Pouvanna A Oopa. In 1953 and 1957 his party won a majority in the territorial assembly, and Pouvanna was elected to the French Senate. However, when Pouvanna began to demand independence, his support dwindled, his party split, and he was abandoned by Protestant Church leaders, whose parishioners amounted to about half the population. His local base weakened, Pouvanna was arrested on what were widely believed to be trumped-up charges and sentenced to jail and exile. Pardoned in 1968, he retained enough of his aura to win election to the French Senate in 1971 and a place in the local pantheon after his death in 1977. But he had ceased to be a significant factor in territorial politics, in which the Gaullist Flosse had become a significant figure.

Pouvanna returned to a Tahiti transformed by government spending for the Mururoa atoll nuclear testing site and a new wave of tourism. The resulting prosperity convinced mainstream political leaders of the advantages of continued association with France. Advocates of independence remained but were never able to win more than 15 percent of the vote. Instead political pressures on France were primarily directed toward securing maximum concessions to local autonomy and maximum French economic support, a campaign in which Flosse began to take the lead in 1980 when he switched from opposing autonomy to demanding it. French receptivity in turn was enhanced by the importance France attributed to its nuclear-weapons capabilities. By the mid-1980s a territorial government with an elected president had been given much greater internal power and more control over coastal and seabed resources, while the authority of the French governor-general was correspondingly much reduced. Tahitian joined French as the territory's official language, the Tahitian flag waved with the *tricouleur* on official buildings, and a new Tahitian national anthem was composed. In 1988 the newly elected president, Alexandre Léontieff, who displaced Flosse for one term, described the territory as having "all the advantages of independence without its inconveniences." "Why independence?" he asked. "We have an autonomous government, the high commissioner of France does not intervene; we are masters of all the economic, social, and cultural decisions."[6]

While the political majority continues to reject independence as an option, pressures for greater autonomy continue. In this context, French testing has been a double-edged sword in the arsenal of the *indépendentistes*. On the one hand it stirs up strong anti-French feelings, as most recently demonstrated by the rioting and demonstrations that greeted the return to testing in 1995 after a three-year suspension. On the other testing has underwritten the prosperity that argues for a continued French connection, while both encouraging French receptivity to the political demands of a strategically important territory and requiring heavy French expenditures there. The French decision to abandon testing after the final explosion of January 1996 was thus a source of some Polynesian anxiety over future support. However France thus far has been reassuring, committing itself to providing $193 million annually for the next ten years to help transform the heavily dependent economy into a more self-supporting one. Meanwhile the continuing drive for increased autonomy and higher status has been embodied in legislation overwhelmingly passed by the territorial assembly late in 1995 and by the French parliament early in 1996. Dismissed by the *indépendentiste* leader, Oscar Temaru, as inadequate, the legislation increases territorial autonomy and also gives French Polynesia some degree of authority over foreign policy by permitting it to sign regional agreements.

Wallis and Futuna

In the territory of Wallis and Futuna, by far the smallest and most isolated of the French territories in the South Pacific, there has been no such concern with political change. In the important role accorded to its three traditional rulers, who share executive power with the French-appointed administrator, it resembles the kingdom of Tonga more than the other French territories. However, its people share with the latter universal suffrage and the right to vote for representatives to the territorial assembly and the French parliament as well as for the French president. In a deeply conservative society with an economy dependent on French subventions and remittances from overseas workers, there is no support for independence.[7]

New Caledonia

Uniquely in the South Pacific, New Caledonia's progress toward autonomy has been marked by anti-colonial violence. The rebellions of 1878 and 1917 were desperate responses to the destruction of the Kanak way of life, not efforts to advance consciously formulated political goals. Only when voting rights were granted in 1946 was the stage set for political action. Among the new political parties, the multiethnic Union Calédonienne (UC) with its motto Two Colors; One People was the most effective in winning Kanak support. Pressing for greater autonomy and increases in Kanak reserve land, the UC in 1953 won 15 of the 25 territorial assembly seats, 9 of them filled by Kanaks.

Briefly the French response seemed positive. In 1956, with the Socialists in power, a *loi cadre* or fundamental law significantly increased the powers of the territorial government and the role of elected officials. In 1957, however, General de Gaulle's newly formed government abrogated the *loi cadre* in deference to the Caldoche fears aroused by UC successes and increased Kanak political power. Independence was one of the options offered in a subsequent referendum. But according to a proclamation attributed to General de Gaulle himself and posted on all public buildings, should New Caledonia have voted for independence, "France will know that you have chosen to leave the nest and she does not expect you to return. She will wish you luck and cease all material and moral aid since you have considered yourself capable … of earning your own way by yourselves."[8] Faced with this prospect, a large majority in New Caledonia voted in favor of continued territorial status.

The repressive measures that followed the abrogation of the *loi cadre* heightened Kanak frustrations. More radical organizations emerged, led by New Caledonia's sixties generation, educated in political activism at French lycées and universities. Among the most active of such groups, the *Foulard Rouges* took its name from the Communard kerchiefs Louise Michel supposedly distributed to the 1878 rebels. Emphasizing conspicuous acts of protest and the Kanak heritage, both traditional and revolutionary, these Kanak parties abandoned UC moderation and attempts to bridge communal differences. They called instead for New Caledonia's independence as the state of Kanaky. Under pressure from the new parties, the UC also abandoned autonomy in favor of independence, remaining the largest Kanak party while losing most of its European support.

Voting rights became a central issue. For the French it seemed a simple matter; universal, equal suffrage was inseparable from French democracy. For the Kanaks however universal suffrage exercised by all French citizens in the country—even those there temporarily on one government assignment or another—was still another way of sidelining the indigenous inhabitants. In the 1970s Kanak opposition to the equal ballot deepened when the nickel boom dashed hopes that population trends would bring back a Kanak majority. Instead another French-encouraged wave of migrant labor, mostly Polynesian, added significantly to the number of those preferring association with France to independence under Kanak rule. Facing another referendum in 1978, Kanak leaders called for confining the franchise to Melanesians and to others with at least one parent born in New Caledonia. When this effort failed, 42 percent of the electorate responded to the Kanak call to boycott the referendum. Of those who voted, 98 percent favored continued territorial status. The depth of the chasm could not have been more strikingly illustrated.

Hardened positions on one side were met by hardened positions on the other. The Kanak Socialist National Liberation Front (FLNKS) organized in 1984 and led by UC president, Jean-Marie Tjibaou, brought most of the Kanak parties together behind the demand for independence, even though differences among them remained quite strong. The Rally for New Caledonia in the Republic (RPCR), led by Jacques Lafleur, became the major but not always united voice of the Caldoche. Continued association with France was supported by some Kanaks (estimated by one authority at 15 to 20 percent)[9] who were opposed to the strong socialist strain in the independence movement or whose civil service or modern sector employment gave them a vested interest in the French presence. Likewise some Caldoche sympathized with the Kanak cause. But on both sides these were the minority, and relations between Kanaks and Caldoche became increasingly bitter and violent.

As tensions heightened and armed struggle became an increasingly ominous prospect, international concern, much of it unfavorable to France, also came into play. New Caledonia's Melanesian neighbors were particularly critical of the French. Polynesians, Australians, and New Zealanders also joined in, although less vociferously. In 1986 they won General Assembly support for a resolution adding New Caledonia to the agenda of the UN Committee on Colonialism.

These external voices had little impact on French policy, however, alterations of which continued to reflect transfers of power in Paris from right to left and left to right. Socialists, Gaullists, and other conservatives agreed that New Caledonia should if possible remain French. But in a pattern unbroken until recent years, conservatives and Socialists disagreed on the impact of concessions to the Kanaks, Socialists regarding them as indispensable to maintaining the French connection, conservatives, like Caldoche, as the beginning of its end. In 1985 with the Socialists again in power, Prime Minister Laurent Fabius devised a plan intended to relieve Kanak fears of perpetual subordination to an alien majority while retaining the principle of universal, equal suffrage and assuring the Caldoche of a continuing role. The Fabius plan divided New Caledonia into four regions, each with significant autonomous powers and its own regional council, whose members together would form the territorial congress. In one region, Noumea and its environs, the Caldoche would clearly dominate, in two, Kanak domination was equally assured; the fourth might go either way. In the territorial congress, however, the Caldoche were likely to dominate, since half the regional assembly seats were allocated to the heavily populated Noumea region.

Elections later in the year resulted in an FLNKS majority in three of the regional councils and an overwhelming RPCR victory in Noumea and its environs. However, in the territorial congress only 16 seats went to the FLNKS against 25 for the RPCR. Moreover, in 1986, when the right returned to power in Paris under Jacques Chirac, the powers and revenues of the regional councils were substantially reduced, while those of the territorial congress were increased.

Violence spurred by Kanak setbacks escalated and climaxed early in 1988 when Kanak extremists in Ouvea killed four gendarmes and kidnapped 20 more who were then liberated by French forces at the cost of 19 Kanak lives. It was now more evident than ever that escalating violence could be averted only by a compromise reasonably satisfactory to both Kanak and Caldoche. After another change in government in Paris, this was achieved by the Matignon Accord of 1988, the product of an intensive joint effort by Socialist Prime Minister Michel Rocard, Jean-Marie Tjibaou, and Jacques Lafleur. Not unlike the Fabius plan, the Matignon Accord divided the country into regions. But these were now three, the Caldoche-dominated Noumea area and two dominantly Kanak areas. While France was to remain responsible for security and defense, government responsibilities otherwise were to fall under the jurisdiction of the territorial congress, the latter exercising a good deal of autonomy. Special provision was made for Kanak welfare and development, including preferences in educational and administrative appointments and budgetary support for rural development, education, and the like. Finally, the accord provided for a referendum in 1998 with the franchise to be confined to those qualified or potentially qualified to vote in 1988.

It would be wrong to say that the Matignon Accord was greeted with great enthusiasm in New Caledonia; debates over its acceptance were quite contentious, particularly among the Kanaks, its critics arguing that it was still another way to postpone independence while subjecting progress to the fluctuations of French politics. Nevertheless, the signing of the accord brought with it a new calm, even optimism. Elections in 1989 passed without incident. The FLNKS and the RPCR both supported Matignon and, unprecedentedly, refrained from attacks on one another during the campaign. Together, they won 72 percent of the votes and 46 of the seats in the 54-seat territorial congress. Even the assassination of the widely venerated Jean-Marie Tjibaou and his deputy in June 1989 did not break the peace. (The crime was committed by a Kanak opponent. Had it been the work of a Caldoche extremist the outcome might well have been different.) In France the Right, capturing the National Assembly in 1993 and the presidency in 1995, has shown no inclination to reverse the process set in motion under the Left.

The approach of the 1998 referendum on New Caledonia's status has put the spotlight on efforts to achieve agreement among the three Matignon signatories—the FLNKS, the RPCR, and France—on the terms of the choice to be offered the electorate. Although pressures for independence have not disappeared, significantly increased French expenditures on education and development in predominantly Kanak areas have evoked a new appreciation of the French connection. Thus in the spring of 1996 FLNKS proposals called for a form of partnership in which a state of Kanaky would enjoy some of the prerogatives of sovereignty while maintaining significant links with France.

The American Connection: Ambiguity Maintained

The United States both remains a colonial power in the South Pacific and regards its overseas dependencies not as colonies but as parts of the national territory whose people enjoy American freedoms and self-government, regard themselves as both Americans and islanders, and are as anxious as Texans or New Yorkers to retain their dual identities. This has been generally true of the peoples of Guam and American Samoa. Of the island groups of the Trust Territory, in contrast, only the Northern Marianas has opted to remain under the U.S. flag. The others—the Federated States of Micronesia, the Republic of the Marshall Islands, and the Republic of Palau—have chosen independence; nonetheless, under the compact of free association, they retain extensive defense relations with the United States and a great deal of U.S. economic support.

Guam and American Samoa

Although both in American Samoa and Guam relations among the three branches of government are based on U.S. balance of power principles, variations in other respects reflect the different role of tradition in the two territories. Guam's constitutional arrangements afford no protection for the traditional institutions that were largely eroded by three centuries of exposure to Spaniards, Americans, and briefly Japanese and by the increasing size of the nonindigenous population, which now outnumbers the indigenous Chamorros. In American Samoa, on the other hand, the United States when its rule began pledged to respect Fa'a Samoa in the compacts with local chiefs. Accordingly, the constitution calls upon the government to protect

> persons of Samoan ancestry against alienation of their lands and the destruction of the Samoan way of life and language.... Such legislation as may be necessary may be enacted to protect the lands, customs, culture, and traditional Samoan family organization of persons of Samoan ancestry.[10]

Members of the upper house of the bicameral legislature are *matai*, but the governor and the members of the House of Representatives are popularly elected. There are also special arrangements for trying cases involving land registration or *matai* titles.

Pressures exist in both territories for fuller participation in the U.S. political system. One source of grievance is the fact that, unlike citizens of French territories, American Samoans and Guamanians cannot vote in U.S. presidential elections, and their representatives in the U.S. Congress are nonvoting also. From the local perspective, the government in Washington retains excessive rights in both legislative and budgetary matters and too often ignores the particular circumstances of the insular territories.

In considering status changes, however, American Samoans weigh heavily the protection of the *matai* system and indigenous land ownership provided by their present arrangements, protection that could be lost under some new status requiring local conformity with U.S. constitutional provisions. Economic considerations also weigh heavily. In 1970 a Samoan congressional committee established to examine questions of political status, while granting that independence might be "the ultimate goal of any freedom-loving people," described it as, for the moment, "economic suicide."[11]

In Guam Chamorro nationalism has become increasingly potent. Pressures for a new arangement are fed by the conviction that Guam, unlike American Samoa and the Commonwealth of the Northern Marianas, is disadvantaged by having come under U.S. sovereignty solely by transfer from Spain without regard to the views of its people. Desires for greater authority over the application of federal law have been especially aroused by concern over Chamorro political status and land rights. Chamorros, already a minority, fear that if immigrants—Filipinos in particular—continue to be admitted and naturalized under U.S. law, their community will lose its hold on political power. Meanwhile, the land issue has become highly contentious as military base drawdowns have released a good deal of U.S.-owned land to which, under federal law, other U.S. agencies have first claim. Arguing against plans to use significant acreage for national wildlife preserves, Chamorros claim that it is they who are an endangered species to whom the land should be restored.

However, given the importance of U.S. financial support, only a small minority even of Chamorros would opt for independence. Alternatives involve problematic trade-offs: a new status bringing Guam under the provisions of the U.S. Constitution, with the electoral rights involved, would deny Guamanians such privileges as the freedom from federal taxes they now enjoy; for its part, Congress is unlikely to look with favor on proposals enabling Guam to override federal laws.

The Trust Territory

Political evolution in Micronesia has been much more complex and difficult than in Guam and American Samoa. For 15 years after World War II the United States largely neglected the 2,000 islands of the Trust Territory of the Pacific Islands, supporting a population of fewer than 150,000 on a land area no greater than half of Rhode Island's. The islands' strategic importance did not survive the war's end, U.S. requirements for forward defense and force projection in this part of the Pacific being largely satisfied by facilities on Guam, fully a U.S. territory and larger than any of the TTPI islands. Accordingly, money appropriated for base construction in seven TTPI sites was never spent. Strategically, it seemed important only to deny military use of the islands to other powers, a task requiring no substantial investment of attention or resources.

Military use of the islands thus reflected not their strategic importance, but their remote location and tiny population, characteristics seemingly making them suitable for activities elsewhere impractical, unacceptable, or overly open to scrutiny. These considerations led to the use for nuclear testing of two minuscule atolls—Bikini and Enewetak—at the westernmost corner of the Marshall Islands. From 1946 until 1958, when testing in the area stopped, there were 67 weapons tests—17 involving hydrogen bombs—with a total explosive power equivalent to 7,000 Hiroshima-type bombs. The problems that would arise in relocating the small number of islanders thought at the time to be affected and in dealing with the aftereffects of the explosions were woefully underestimated. Even in the mid-1990s, earlier claims settlements are being brought into question as formerly classified information is disclosed and new studies reveal more about the range and long-term impact of radioactive fallout. With less fateful consequences, remoteness and the ability to exclude outsiders were also key factors in the selection of Saipan as a secret training site for Chinese Nationalist troops and for the continuing use of Kwajalein atoll as a splashdown site for unarmed intercontinental ballistic missiles test fired from California.

Nuclear testing aside, with no occasion for a large-scale U.S. military presence in the Trust Territory and little to inspire other interests, policymakers in the 1940s and 1950s thought it best to leave the Micronesians undisturbed in their traditional way of life. The traditional way of life, however, had already been much disturbed. Japan's policies, although intended for its benefit and those of its settlers, had provided roads, communications, and various modern amenities to which the islanders also had access and which had already altered traditional patterns. The war had destroyed this infrastructure and much else; the skills to restore the prewar economy and much of the will to do so disappeared when the Japanese settlers were repatriated. In the absence of resources and external help for rebuilding further deterioration was inevitable; in 15 years or so the Trust Territory had become the Rust Territory.

In 1960 and 1961, however, domestic and international developments brought U.S. inattention to an abrupt end. Representatives of Hawaii, a state since 1959, were calling for extending federal programs to the Trust Territory. Micronesians meanwhile were observing with interest progress toward independence in Western Samoa and political developments in other parts of the island world. In 1961 U.S. policy was directly challenged when a UN visiting mission severely criticized almost every aspect of the Trust Territory's administration. The recently inaugurated Kennedy administration responded rapidly to this blow to the American self-image. Under its auspices and those of the Johnson administration numerous federal programs were extended to the islands and the Peace Corps arrived en masse. Political development, hitherto confined to the local level, was extended to the territorial administration with the establishment in 1964 of the popularly elected bicameral Congress of Micronesia. However, extensive authority remained in the hands of the high commissioner and the secretary of the interior.

Extensive as were its programs, the Kennedy administration gave little thought to the TTPI's future political status; it assumed that large investments of money, personnel, and goodwill, providing local opportunities for participation and prosperity, would eventually convince the Micronesians to opt for continued association with the United States. However, it soon became evident that for Micronesian leaders status was a preeminent concern.

In 1967, having received no response to its request for a joint study of the status question, the Congress of Micronesia established its own Future Political Status Commission under the chairmanship of Lazarus Salii of Palau. The commission's report, presented in 1969 and approved by the Congress of Micronesia, rejected independence. It called instead for a new partnership meeting Micronesian requirements for continued protection and economic assistance while providing the United States with continued access to Micronesian territory for strategic purposes. But any such partnership, the report read, must respect "our wish to live as Micronesians, to maintain our Micronesian identity, to create a Micronesian state" and recognize "that the basic ownership of these islands rests with Micronesians and so does the basic responsibility for governing them."[12] The Micronesians, in addition, were to have the right to draft and amend their constitution and to terminate the U.S. relationship unilaterally.

Although all of these principles would be embodied in agreements reached 20 years later, the Nixon administration met them with a strikingly different proposal calling for converting the Trust Territory into an unincorporated territory of the United States. The United States would continue to exercise veto powers and the right of eminent domain, while the Micronesians would have no authority to terminate the relationship. Lazarus Salii expressed the depth of Micronesian bitterness. Micronesia, he said, "would become the newest, the smallest, the remotest nonwhite minority in the United States political family—as permanent and as American, shall we say, as the American Indian."[13] In this ominous manner the pattern was set for negotiations that were to prove extraordinarily prolonged and exasperating to both sides.

Differences over the proper objectives of the negotiations remained substantial for some time. In essence the Micronesians wanted both maximum sovereignty and maximum U.S. economic support; the United States wanted to preserve its right to military use of Micronesian territory to the extent future contingencies might require. Although accepting the necessity of a continued major financial contribution to the islands, the United States also wanted to be able to ensure its proper utilization and accounting. Combined with these differences in principle, efforts to make detailed arrangements on a wide variety of subjects fed the litigious instincts of both sides. Prolonged negotiations provided ample opportunity for seemingly settled problems to reemerge in new forms and for new issues to surface.

The labyrinthine U.S. policy-making process slowed the negotiating pace. Decisions could be made in London, Canberra, and Wellington by governments confident their parliamentary majorities would approve. In Washington the executive branch could have no such confidence, since congressional authority over territorial matters was jealously guarded against executive encroachment by Senate and House Interior Committees. However, although congressional views could be decisive, it was not always easy to determine what they were or even to attract congressional attention.

Nor was the executive branch itself single-minded in its attention to Micronesian issues. Changes in administration rarely brought sharp French-type reversals of policy. But they did require a review of the bidding which, in matters of lesser urgency, came very slowly. Among the large number of executive agencies involved, three were central to the process: Interior with its administering responsibilities; State, where responsibility was divided between officers concerned with regional policy and those concerned with UN affairs; and Defense, with its internal overlapping and competing jurisdictions. The coordinating functions of the National Security Council gave it a role to play when it chose to do so. Other agencies also became involved in questions concerning the future of federal programs.

Each of the principal departments tended to blame the others for lack of progress. State was seen as slow to reach internal agreement, too anxious to conciliate the UN, and given to clandestine efforts to supplant Interior's authority. Defense was seen as rigidly bound by the contingency thinking that made it unwilling to give up any currently held position or privilege. Interior was seen as inflexible, parochial, and too preoccupied with retaining congressional favor. And, while the executive branch moved slowly and with difficulty, interested members of Congress and their staffs were easily convinced that the process was moving too rapidly and without adequate attention to congressional views. In 1970, in an effort to bring about greater order and cohesion, a new mechanism was placed at the top, the Office of the Micronesian Status Negotiations, headed by a special representative of the president with ambassadorial rank. However, the additional ingredient had little of the precipitating effect intended.

The Micronesians themselves found it increasingly difficult to negotiate with a single voice. The single jurisdiction imposed by foreign sovereigns had no basis in indigenous traditions or relations, while the passage of time and increasing local autonomy sharpened the differences among the six U.S.-established districts—the Mariana Islands, the Marshall Islands, Palau, Ponape (now Pohnpei), Truk (now Chuuk), and Yap. In the Marianas, the most westernized of the Micronesian islands, separatist tendencies became evident quite early, encouraged by the relative prosperity resulting from Saipan's role as TTPI headquarters. Wanting to preserve this prosperity through the close association with the United States enjoyed by their fellow Chamorros of Guam, Marianas leaders were also convinced that other Micronesians had designs on their relative wealth. Early in 1971, incensed when the Congress of Micronesia passed an allegedly discriminatory income tax law, the Marianas district legislature agreed unanimously that the district should secede "by force of arms if necessary and with or without the approval of the United Nations."[14] The rapidity with which the United States abandoned its earlier commitment to Micronesian unity and the speedy achievement of agreement on the specifics of the territorial arrangement prompted charges of U.S. involvement in the decision to separate. However, while the U.S. military in particular certainly welcomed the Marianas decision to opt for territorial status, Marianas separatist sentiments were equally strong. In a 1975 UN-observed plebiscite 79 percent of those voting (98 percent of the electorate) supported the commonwealth arrangement.

Under a constitution of its choosing, the Commonwealth of the Northern Marianas is internally self-governing with a popularly elected governor and bicameral legislature. The federal government, which remains responsible for foreign policy and defense, was accorded a 50-year renewable lease on land amounting to some 15 percent of the total and the same right of eminent domain it exercises in the states. Otherwise land ownership and leases beyond 55 years are limited to those of Northern Marianas descent.

Inequality among districts was also a divisive factor elsewhere in the Trust Territory. The Marshalls and Palau complained bitterly about the larger share of appropriated funds granted to the poorer but more populous Truk and Ponape. Outvoted in the Congress of Micronesia, Marshall Island leader Amata Kabua described it as "scheming to control our resources, steal our revenues, and jeopardize our right to self-determination."[15] Prospects in the Marshalls for increased revenues from the Kwajalein missile site and in Palau for a large Japanese oil-storage facility only increased their resistance to sharing their resources with their neighbors. The die was cast in 1978 when the voters of Palau and the Marshalls rejected a constitution drafted under the auspices of the Congress of Micronesia. The Congress of Micronesia was dissolved; Yap, Truk, Kosrae, and Ponape then constituted themselves the Federated States of Micronesia (FSM); and the United States was faced with the prospect of negotiating with three separate teams.

By this time, however, prospects for more rapid movement had improved in other respects. The United States had accepted principles central to the Micronesians that it had earlier rejected: free association as an appropriate outcome; the right of unilateral termination by either side; Micronesian authority over foreign affairs to the extent consistent with U.S. responsibility for security and defense; and Micronesian authority over offshore marine resources. In 1980 a compact of free association embodying these principles was initialed by the Marshalls, the FSM, and Palau. Two years passed before the compact was signed, with some of the delay occasioned by a new administration in Washington and some by continued textual changes. More time-consuming steps remained, however: UN-observed plebiscites; passage by the U.S. Congress and approval by the president of legislation embodying the compact; and dissolution of the trusteeship by the Security Council.

Pending Security Council action, which did not take place until 1990 after the end of the cold war had removed the prospect of a Soviet veto, the compact was put into effect in the FSM and the Marshalls in 1986 by presidential order. In Palau, however, the process had been stalled several years earlier by a conflict between Palau's nuclear-free constitution and compact provisions for U.S. military access and transit rights. The conflict was not resolved until October 1994 but in the interim, for most practical purposes, Palau functioned as a freely associated state under the terms of the compact.

Under the compact, the three freely associated states are fully sovereign internally. With the support of a plebiscite, each may terminate its relationship with the United States. Their foreign affairs authority is limited only by the requirement for consistency with U.S. defense responsibilities. The compact and related agreements provide for continuing large U.S. subventions in the form of grant aid, federal programs, and gratis federal services. Its defense commitment gives the United States the authority to disapprove acts compromising U.S. security and to exclude third-party military activities. Separate agreements establish the conditions under which the United States may use land for defense purposes. Still another agreement sets up an interest-bearing $150 million endowment fund to settle the claims of Marshall Islanders affected by U.S. nuclear testing. In the Marshalls and the FSM the aid provisions and some of those relating to defense and security will expire in 2001; negotiations on such future arrangements are to begin in 1999 and will almost certainly result in a reduced financial commitment. In Palau the aid provisions will expire in 2009; military provisions run until 2044.

Like other island polities, the associated states have adopted constitutions that mirror those of Western democracies in their provisions for representative government, universal suffrage, judicial independence, and guaranteed rights. The FSM and Palau are federal states in which separation of powers prevails; the Marshalls has adopted the parliamentary system.

All three constitutions pay deference to custom. The constitutions of Palau and the FSM protect the "role and function" of traditional leaders "as recognized by custom and tradition." The FSM constitution provides that if a statute intended to protect tradition is challenged as violating fundamental rights, the protection of tradition is to be considered "a compelling social purpose warranting such governmental action." In the Marshalls, 12 paramount chiefs make up the Council of Irois, which advises the cabinet and may request the legislature to reconsider any bill affecting customary law. Customary land rights are constitutionally protected and land alienation is severely restricted.[16]

*   *   *

In the 50 years since the end of World War II, independence has come to 14 island states. While valuing their sovereignty, all would agree that independence is a mixed blessing, involving as it does efforts to cope with the multitudinous problems of modernization without, in most cases, the resources and potential that might provide freedom from continued heavy dependence on the economic support of former metropoles.

Notes

1. Fiji was expelled from the Commonwealth during the coup year, 1967.

2. Yash Gai, ed., *Law, Government, and Politics in the Pacific Island States* (Suva: University of the South Pacific, 1988), pp. 36, 37.

3. Ibid., p. 214.

4. Davidson, *Samoa Mo Samoa*, p. 410.

5. Douglas Oliver, *Black Islanders* (Honolulu: University of Hawaii Press, 1991), p. 186.

6. *Pacific Islands Monthly*, March 1988.

7. There are some 13,000 Wallisions in New Caledonia.

8. Crocombe and Ali, *Foreign Forces*, p. 7.

9. Albert B. Robillard, *Social Change in the Pacific Islands* (London: Kegan Paul International, 1992), p. 76.

10. Stanley A. de Smith, *Microstates and Micronesia* (New York: New York University Press, 1970), p. 111.

11. Van Cleve, *Office of Territorial Affairs*, p. 79.

12. Donald F. McHenry, *Micronesia: Trust Betrayed* (Washington, D.C.: Carnegie Endowment for International Peace, 1975), pp. 92, 94.

13. Vincente T. Blaz and Samuel S.H. Lee, "The Cross of Micronesia," *Naval War College Review*, June 1971, p. 83.

14. Ibid., p. 81.

15. Grant K. Goodman and Felix Moos, eds., *The United States and Japan in the Western Pacific* (Boulder, Colo.: Westview Press, 1981), p. 86.

16. Yash Gai, *Law, Government, and Politics*, p. 41.

PART THREE

The Practice of Politics

6

Running the State

Neither Paradise nor Paradise Lost, the Pacific island countries are faced with problems of governance unique less in nature than in scale. Their governments, like many, are challenged by poverty, crime, corruption, youth anomie, drug abuse, population pressure, slow growth rates, resource depletion, environmental degradation, and intermittent natural catastrophes. Other problems peculiar to their circumstances burden individual governments. These include extreme localism and a particularly high rate of crimes of violence in Papua New Guinea and, in a number of countries, fissures between major segments of the population—indigenes and Indians in Fiji, Caldoche and Kanaks in New Caledonia, and Anglophones and Francophones in Vanuatu. Indigenous resources for coping with these problems are extraordinarily limited. Most of the island countries are not yet capable of self-support. Some probably never will be. All except Nauru still receive foreign aid, which provides essential support but maintains dependence and puts pressures on traditional ways without necessarily providing sustainable alternatives. Nevertheless, among developing countries the Pacific island polities are far from the bottom of the welfare scale. A United Nations Development Program (UNDP) report comments, "The quality of life and human welfare in the Pacific is enviable in comparison with many of the developing countries of Africa, Asia, and Latin America." In addition to foreign aid the UNDP identified as contributing factors a large subsistence sector, strong cultural identity, stable social fabric, and strong government support for "people-oriented" development.[1]

Pacific island political leaders, no more immune from self-interest, corruption, and divisiveness than their counterparts in other places, no more capable of providing complete and lasting solutions, have nevertheless shown considerable ability to cope with the problems of governance. Parliamentary systems function, even though they are unsupported in most cases by strong party systems, and elections are free and take place regularly. Peaceful, orderly, constitutional succession has been the general rule, with Fiji the major exception. The churches continue to play a generally accepted political role, while ideology plays practically none. Only in Fiji has the postcolonial constitution been displaced by undemocratic means. Only in Papua New Guinea have demands for change been pursued amid protracted violence.

Tradition and Modernization

In politics, as in other spheres of life, pressures to change and pressures to maintain traditional ways have existed side by side. In conservative societies, where individual and group identification is closely tied to tradition and custom, the adjustment has not been easy. Jean-Marie Tjibaou, westernized by his training as a Jesuit priest but also deeply attached to New Caledonia's Kanak tradition, defined the dilemma: "How to come to terms with this enormous change and yet remain ourselves while feeling comfortable in this new environment... ? Our struggle now is about being able to include as many characteristics as possible of our own past, our own cultures, in the human and social model we want to pursue."[2] Few would disagree with Tjibaou's definition; it is in the implementation that angry disputes emerge.

In New Caledonia, the Kanak-oriented French development program has stimulated dissension between modernizers and traditionalists who argue that the French are seeking only to impose alien capitalist institutions. Thus, one longtime Kanak nationalist leader, Nidoish Naissaline, accuses France of wanting "to make us believe that it is our traditions which impede the development of the territory and the Kanak people, that our traditions are bad and need to be eliminated in order to progress." Denouncing Kanak politicians as collaborationists, he urges Kanak chiefs to act against those who want to sabotage custom and claims, "Either it's the high chief or the political party commands." Expressing the other side of the debate, Tjibaou's successor as Union Calédonienne president asked, "Are we working for the future of a Kanak republic or a multitude of hereditary kingdoms?"[3]

Throughout the region the role of hereditary leaders or chiefs has continued to evolve, although not in consistent ways. In Western Samoa a 1990 referendum victory for universal suffrage reflected the discredit that had fallen upon the *matai* system as titles proliferated from 4,700 in 1961 to 20,600 in 1988. *Matai*, it appeared to many, were no longer the bulwark of Fa'a Samoa. Instead, as villages divided old titles and created new ones to increase their electoral influence, they became "pawns in the political game," manipulated by people seeking the perquisites of parliamentary membership.[4] The *matai* system nevertheless remains politically important.

The 1990 referendum result did not express a strong popular demand for universal suffrage; a majority of the electorate either abstained or voted no. Moreover, candidacy is still limited to holders of *matai* titles—between one-fifth and one-quarter of an electorate of 85,000 to 90,000 people.

In Tonga also, after more than a century of little change, the sense that traditional institutions have become too corrupt to serve their original purpose has fueled the aspirations of an emerging business and professional class. The typically gradualist advocates of reform profess nothing but respect for the widely venerated, traditional monarchy. Nevertheless, their pro-democracy stance envisages an end to the power monopoly of the king and the hereditary nobles, who hold two-thirds of the parliamentary seats. They urge instead cabinet responsibility to a popularly elected parliament and greater government transparency.

Emerging as an active force in the late 1980s, the pro-democracy movement, led by former teacher 'Akilisi Pohiva, has been strongly supported by the leadership of the established Free Wesleyan Church and the Roman Catholic hierarchy. In 1990 a more coherent and widely supported protest movement emerged, precipitated by the government's sale of passports to foreigners (mainly Hong Kong Chinese) and its refusal to account for the resulting huge revenues. Despite electoral successes in 1993, however, the reformers have been unable to unite around a common program and organization, and their royal and noble opponents retain their grip on power. Although the crown prince of Tonga is believed to accept the inevitability of change more readily than the septuagenarian king, he does not seem anxious for it to come soon. Democracy, he has said, "has to evolve eventually."[5] Even so, the movement seems to have had some impact on Tongan government operations. According to one observer, the government is revealing more about its activities, ministers are more willing to account for their actions and remain in the country during parliamentary sessions, and steps are being taken to reactivate local government.[6]

Elsewhere, in contrast, there are pressures for strengthening chiefly roles, sometimes in connection with movements for greater local autonomy. Thus a Solomon Islands constitutional review committee in January 1988 set forth two decentralization options, both giving new powers to traditional chiefs. Submissions to a similar review committee in Vanuatu also looked to enhancing the power of the councils of chiefs established by the constitution at district, island, and national levels. Supporting this position, the review committee's vice chairman observed: "The constitution needs to include legal recognition of customary rights and organizations such as the Malfatumauri (National Council of Chiefs). They need to be recognized as the authentic power."[7]

Political Leaders and Their Parties

Whether of chiefly or common rank, island leaders, mostly quite young when they attained political eminence, have not quickly faded from view. In line with the Westminster system, fallen prime ministers remain active as leaders of the opposition. But the prevalence of weak party systems and coalition governments also means that yesterday's prime minister may turn up tomorrow as a member of his rival's cabinet. In the Solomon Islands, for example, Sir Peter Kenilorea became the country's first postindependence prime minister in 1978, leader of the opposition in 1981, and prime minister once again in 1984. In 1986, defeated in the parliamentary vote for prime minister, he later became deputy prime minister to his principal rival.

Only in Vanuatu and Fiji have political leaders been supported for significant periods by well-organized political parties normally enjoying parliamentary majorities; in both cases, these parties have disintegrated under the political pressures of recent years. Only in New Caledonia and Fiji have parties been strongly differentiated by long-standing policy differences—in New Caledonia, independence or continued association with France; in Fiji, Fijian dominance versus ethnic equality. Elsewhere policy differences run deepest where the issues involved pit local interests against national ones.

In some of the microstates parliaments of 30 or fewer members do not provide much support even for the existence of parties. In Tuvalu, for example, which has no parties, governments are formed at secret sessions of the 12-member parliament when those with compatible interests choose a leader among themselves and assign ministerial responsibilities. Even in quite tiny countries, however, political parties may spring up in response to issues of the moment. In Niue dissatisfaction with Sir Robert Rex's 16-year rule led to the organization of the Niue National People's Action Party (NPAP). The new party, however, was successful neither in the 1991 parliamentary election nor in the no-confidence vote it sought to organize some months thereafter. Meanwhile, the future of the NPAP's role as an opposition party was cast in doubt when Young Vivian, its very popular founder and leader, accepted a seat in the Rex cabinet. In the general view Vivian had abandoned his role as the prime minister's principal opponent, becoming instead his heir apparent, an expectation fulfilled a year later when Rex died at the age of 84 and Vivian succeeded him.

Typically, the weak and sometimes short-lived parties of the South Pacific are most active in mobilizing the vote during election campaigns, unimportant in determining the composition of the government or its policies and primarily concerned with advancing local or clan interests. Even as electoral machines, they are frequently ineffective. In Papua New Guinea in 1987, although 14 parties contested the general elections, two-thirds of the candidates were independents who together won 40 percent of the votes.

In the postelection bargaining over the composition of a new government, party loyalty and support play a lesser part than personal ambitions and strengths. Not infrequently, when a newly elected parliament first convenes, the political affiliations of some of its members are unknown and prime ministers may be chosen by the slimmest majorities. In Papua New Guinea, for example, Paius Wingti became prime minister by a one-vote majority in 1992, as did the Solomon Islands' Francis Billy Hilly in 1993. Both were able subsequently to build up their majorities and remain in office until 1994.

Coalitions are easy to achieve. They are unobstructed by rigid differences of principle, sharply conflicting approaches to the role of the state, or strong policy attachments. Normally, they come together after elections or votes of no confidence and comprise an assortment of leaders who together command a parliamentary majority and who are able to agree on who among themselves should be prime minister. Competition for this post may be as keen within parties as between them; to be acknowledged as head of the leading party is not a guaranteed road to leading the government. Key considerations for those seeking membership in the ministry are the powers and perquisites cabinet office brings; these in turn encourage disproportionately large cabinets. Thus, of the 47 members of parliament elected in the November 1994 Solomon Islands general elections, 18 became cabinet members. Where parties are predominantly the voice of regional loyalties, geographical distribution can be an important factor in coalition building.

Coalitions are easily unseated between elections. A handful of members of parliament by switching loose allegiances can bring about a vote of no confidence and eventually, after much maneuvering and bargaining, a new government, often containing a number of holdovers from the preceding cabinet. Until PNG's constitution was amended in 1992, prime ministers, lacking the countervailing power to dissolve parliament, were particularly vulnerable to being unseated by no-confidence votes, which could be moved at any time after the prime minister's first six months in office. However, a no-confidence vote during the last six months of a parliamentary term automatically dissolved parliament, thereby precipitating general elections. In 1992 a long-debated constitutional amendment extended both periods—the prime minister's immunity to 18 months; the automatic dissolution period to 12. The following year, 15 months into his term, Prime Minister Wingti dissolved his cabinet and obtained parliamentary approval, thereby claiming immunity from no-confidence votes for another 18 months. Wingti later stepped down after his actions were declared unconstitutional by the Supreme Court and was succeeded by Sir Julius Chan, deputy prime minister and foreign minister in the Wingti cabinet.

Leaders cannot rely on party discipline and loyalty to prevent party members from crossing the aisle in response to the inducements offered by their rivals. At the same time, however, the weaknesses of political parties may provide a leader with welcome opportunities to play his own hand. In the Solomon Islands, for example, in 1990 Prime Minister Solomon Mamaloni forestalled an anticipated effort by his own People's Action Party (PAP) to unseat him simply by firing the five party cabinet ministers and replacing them with five members of the opposition.

Corruption and Its Consequences

The fall of governments and political leaders often involves corruption charges. However, the line between corruption and the redistributionist demands of custom has sometimes been difficult to draw. Arguing against restrictions on the business activities of government members in the PNG leadership code, longtime politician Iambakey Okuk argued that the accumulation of wealth is an essential element of political status in Melanesia.[8] When the Western Samoan Supreme Court deprived former prime minister Tupua Tamasesie Efi of his parliamentary seat on vote-buying charges, the defense argued that, as holder of one of the country's highest titles, Efi was expected to give gifts in the villages he visited. The court ruled, however, that in this instance the provisions of the electoral law superseded custom.[9]

As this court decision illustrates, although the requirements of custom certainly play a part in the motives and rationalizations of the corrupt, tradition is far from providing an unchallenged justification. There is no lack of laws defining in wholly nontraditional terms corrupt practices and the penalties for engaging in them. Courts and review boards for the most part examine the cases brought before them actively and independently. Expulsion from office on corruption charges, however, is not necessarily a bar to later return.

External Ties and Ideology

The parties of the U.S. and French dependent territories are unique in their ties with parties overseas. In American Samoa and Guam the principal political parties are Democratic and Republican branches with a voice, albeit very small, in the affairs of their national organizations. In the French dependencies, where territorial citizens vote in French presidential and national assembly elections, the metropolitan parties have a direct stake in alliance with territorial ones. The latter, in turn, look to help from metropolitan parties of compatible views. In New Caledonia, Kanak parties tend to throw their support behind the Socialists, Caldoche parties behind right-wing parties, including the far right National Front of Jean-Marie Le Pen. The French Right is expected to do what it can for the Caldoche, the Left for the Kanaks. In French Polynesia the party Gaston Flosse leads is a branch of the neo-Gaullist Rassemblement pour la République; the close ties he has maintained with the French Right have helped support his claims to office at home and elevated him for two years to a junior cabinet post in Paris. In 1992, when the French high commissioner took control of a seriously unbalanced French Polynesian budget, Flosse, attributing the problem to mismanagement by his predecessor, accused the Socialists, then in charge in Paris, of deliberately delaying action until after he had won the presidency back from their protégé, Léontieff.

Parties in the U.S. and French dependencies may find philosophic affinities as well as practical advantages in their external ties. The ideological battles of the cold war, however, played virtually no part in island politics, the apprehensions of Western powers to the contrary notwithstanding. These were aroused in the mid-1970s by a newly manifested Soviet interest in the region, the fervor of the South Pacific anti-nuclear movement, and the conspicuous role of Australian and New Zealand far left activists in pacifist and trade union organizations. They were reinforced in the mid-1980s when a number of Kanak militants and members of the Melanesian resistance in Irian Jaya received military training in Libya. Vanuatu's foreign policy also aroused concern. Under Father Lini's Vanuaaku Pati government, bitterness against the French seemed to spill over into antagonism toward the West as a whole. Vanuatu became the only South Pacific member of the Non-Aligned Movement, frequently joined in radical third world condemnation of one or another Western power, and tilted its diplomatic relations toward the Soviet bloc. In 1986 Western apprehensions were distinctly heightened by the news that the group of young Vanuatu citizens who had gone to Libya had been sponsored by Father Lini's closest associate and deputy, Barak Sope, noted both for his business success and his anti-Western sentiments.

However, the prolonged political turmoil that followed had almost nothing to do with ideology and almost everything to do with a bitter power struggle in the VP, starting with conflict between Lini and Sope and becoming much broader. Issues of substance were quickly lost to sight—Sope's breach of VP discipline in unilaterally making the Libyan contact and his claim to represent traditional land owners, whose rights he alleged were being undermined by Lini's government. Lini's effort to retain power was undermined by his increasingly arbitrary and erratic behavior in the wake of a stroke in 1987 and a heart attack in 1991. A three-way split finally resulted with Lini, Sope, and former VP stalwart Donald Kalpokas each leading a new political party. Although seeming to totter on occasion, the constitutional system was preserved. But the Vanuatu political scene was transformed. Prolonged divisions in the formerly disciplined and cohesive VP had given the Francophones, hitherto a fringe element in national politics, opportunities for maneuver and alliance building. In the 1991 elections, Maxime Carlot's Francophone Union of Moderate Parties won 19 seats, against 10 each for Lini and Kalpokas and only 4 for Sope. Vanuatu thus joined its Southwest Pacific neighbors in government by coalition. Its foreign policy also became distinguishable from that of its neighbors only by its more outspoken support for a continued French role in the region.

Religion and Politics

The absorption of Christianity into the South Pacific tradition is as important in politics as in other aspects of life. Churches, clergy, and lay religious leaders not only wield great influence and enjoy considerable prestige; priests, pastors, and lay leaders also actively participate in the political competition as party leaders and officeholders. In Tonga the campaign for a more democratic system has led to a virtual war of words between church and state. Attacking Wesleyan and Roman Catholic religious leaders for supporting Pohiva's reform movement, the speaker of Tonga's parliament, himself a noble, dismissed them as no more than power seekers, convinced "that their empires are not big enough."[10] The senior pastor of the established Free Wesleyan Church, headed by the king, charged the constitution with allowing "a despot to be a despot" and described constitutional codification of inequality as "ethically criminal and theologically barbaric."[11] In scarcely more measured terms, Tonga's Catholic bishop denounced the king's reported plan to organize a political party to counter Pohiva's movement as a subterfuge produced by "the devious and unenlightened."[12]

Notes

1. United Nations Development Program, *Pacific Human Development Report* (Suva: UNDP, 1994), pp. 6, 7.

2. Michael Spencer, Alan Ward, and John Connell, eds., *New Caledonia: Essays in Nationalism and Diversity* (St. Lucia: University of Queensland Press, 1988), p. 65.

3. Donna Winslow, "Sustainable Development in New Caledonia," *Pacific Affairs*, Winter 1991–1992), pp. 501–502.

4. Yash Gai, *Law, Government, and Politics*, p. 75.

5. *Pacific Islands Monthly*, May 1993, p. 36.

6. Kerry E. James, "Tonga's Pro-Democracy Movement," *Pacific Affairs*, Summer

1994, pp. 243–259.

7. *Pacific Islands Monthly*, July 1991, p. 23.

8. May and Nelson, *Melanesia*, p. 644.

9. Foreign Broadcast Information Service, *Daily Report: East Asia* (hereafter FBIS-EAS), November 25, 1991.

10. Ibid., April 18, 1991.

11. Ibid., November 24, 1992.

12. Ibid., November 12, 1992.

7

Challenges to Constitutional Order

Only in Fiji and Papua New Guinea has there been a serious challenge to the legitimacy of the political institutions established at independence—in Fiji, to the constitution itself; in Papua New Guinea, to the continued existence within its postindependence borders of a single national state. Developments in these relatively large, politically important, and resource-rich Pacific island states have attracted more international attention than is normally focused on the region. In both, the claims of indigenous people to special rights, particularly with respect to land, have provided the spark. Throughout the islands indigenous land rights can come into conflict with economic development and environmental pressures. But in Fiji and Papua New Guinea they have combined with unique local circumstances to bring about unusually painful confrontations.

In Fiji, constitutional government was displaced entirely: for five years Fiji was ruled by self-appointed leaders whose ranks did not include Indians. A new constitutional order, reinforcing Fijian primacy while not depriving Indians of a role, has brought Indian parties back to the political forum but has not reconciled their leaders to their subordinate role, nor fully satisfied Fijian demands for unassailable protection.

In Papua New Guinea, the struggle between Bougainville militants and the central government continues. It has brought great suffering to the province and exacerbated two of PNG's endemic problems—lawlessness and armed forces indiscipline. But the constitutional system remains intact and the Port Moresby game of politics goes on as usual.

Developments in both countries have had external repercussions. Two military coups, the suspension of the constitution, and the assault on Indian rights exploded Fiji's somewhat exaggerated reputation as the third world's model multiethnic democracy. Fiji was expelled from the Commonwealth, its relations with India were broken, and those with its most important partners—Australia and New Zealand—deteriorated markedly. Relations with India have not been resumed. However, it was not long before the Western democracies, the United States as well as Australia and New Zealand, accommodated themselves to a new Fiji. Its leaders, after all, were familiar figures who, facing no active resistance, seemed to be moving toward revived constitutionalism in a reasonably orderly and moderate manner and who in any case were not about to change their course at foreign insistence.

The Bougainville problem has had more dramatic external repercussions, deeply disturbing PNG relations with the nearby, ethnically related Solomon Islands. Port Moresby has been angered and frustrated by the rebels' ability to use Solomon Islands territory as a safe haven, supply base, and advocacy center for their cause. Honiara, which disclaims any official involvement, has been affronted by seizures in its waters, cross-border raids, and casualties among its citizens.

Fiji

Until 1987 Fiji's 1970 constitutional system seemed to be working as intended to Fijian advantage. Chiefly status remained an important component of leadership. Ratu Sir George Cakobau, the first Fijian governor-general, was the great-grandson of the Cakobau who signed the deed of cession. He was among the country's highest chiefs, as were his successor, Ratu Sir Penaia Ganelau, and both the long-serving prime minister, Ratu Sir Kamisese Mara, and his wife, Adi Lady Lala.

Ratu Mara's Alliance—a Fijian-dominated coalition of three ethnic parties—had assumed office before independence and thereafter maintained an only briefly interrupted hold on power. Its Fijian component was preeminent in its community. Its Indian component represented those members of the Indian community who accepted the need to recognize that, as one leader put it, "the Fijians feel their security lies in Fiji being in Fijian hands politically" and that, like the Malays in Malaysia, "they are prepared to share political power but not to relinquish it to others."[1] A minority in its community, the Alliance normally won 15 to 24 percent of the Indian vote. The third component represented the General Electors, citizens of European or mixed blood.

The Indian-dominated opposition, led by the National Federation Party (NFP), was far from united. The NFP suffered from intense leadership rivalries; Hindus and Muslims were often at odds; and the interests of a prosperous, free-migrant-descended business class were frequently opposed to those of the descendants of the contract laborers, whose trade unions formed the backbone of the NFP. In competition with the badly divided Indians, the relative cohesion of the Fijians and their well-established authority structure gave them an advantage that was reinforced by the support they could expect from the General Electors.

Elections and Coups

The general elections of April 1977 gave warning that Fijian rifts could nullify the safeguards of the electoral system. The newly organized Fijian Nationalist Party, demanding Indian repatriation and "Fiji for the Fijians," captured 24.4 percent of the Fijian vote. Its share of the Fijian vote shrinking from 83 percent in 1972 to only 67 percent in 1977, the Alliance lost its majority, winning only 24 seats. The NFP won 26 seats, its share of the Indian vote having risen markedly owing to resentment throughout the Indian community caused by the government's discriminatory university admissions. However, if Fijian differences had caused the Alliance defeat, Indian differences saved the Alliance day. Citing the NFP's inability to agree on a prime minister, the governor-general named Mara prime minister in a minority government. In September Mara's government fell to an NFP-organized no-confidence vote. In a second 1977 election, a sobered Fijian community returned Ratu Mara to power with 36 seats in Alliance hands.

Thus restored, the foundations of the political accommodation still left ample space for festering fears and resentments. Although a majority, the Indians remained politically disadvantaged, their great economic success shadowed by permanent exclusion from landownership and the uncertainties of leaseholder tenure. Fijian concerns over the size of the Indian community were not wholly dispelled by population-trend forecasts of a Fijian majority in the near future. Meanwhile, despite their ownership of 80 percent of the land, Fijians came second economically to the Indians, who dominated retail trade, were heavily represented in the professional and public service ranks, and on leased land produced 90 percent of Fiji's principal export crop—sugar.

The 1982 general elections brought Fijian fears closer to the surface. Although the Alliance prevailed, it lost 4 seats, holding 28 to the NFP's 24. The loss, which reflected a significant decline in votes for the Alliance's Indian component, prompted accusations of Indian betrayal. In parliament, in the Great Council of Chiefs, and elsewhere, there were demands for constitutional amendments to ensure Fijians a permanent parliamentary majority and sole access to the positions of governor-general and prime minister.

Five years later, in the April 1987 elections, the Alliance lost its majority. A divided Fijian vote reflected both regional and class issues. A new party, the Western United Front, was organized by chiefs of the western provinces, producers of most of Fiji's wealth, who resented their subordination to the political authority of the eastern chiefly establishment—represented at its apex by Ganilau and Mara. In an electoral alliance with the NFP, the Front took 7 percent of the Fijian communal vote. Meanwhile, dissatisfaction with the status quo had attracted urban professional and working-class Fijians to a new multiethnic party, the Fijian Labor Party (FLP), which had also won the allegiance of Indian politicians disillusioned with NFP leadership squabbles. The new party was led by a western Fijian commoner, Dr. Timoci Bavadra, the widely respected president of the Fiji Public Servants Association; its secretary-general, Krishna Dutt, was president of the Fiji Teachers Union. Describing itself as "democratic socialist," the FLP professed respect for the chiefly system but also called for the reduction of chiefly prerogatives. In the election the FLP and the NFP in coalition won 9 percent of the Fijian vote and captured 28 seats to the Alliance's 24.

The outcome—Bavadra as prime minister and a coalition cabinet with Indian ministers equaling Fijian supported by a parliamentary majority including only seven Fijians—prompted apocalyptic visions of the future. The government's assurances that its policies would rest "fundamentally on its recognition of Fijian rights and interests as enshrined in the Constitution of Fiji" were outweighed by primordial fears that Fijians would be reduced in their own country to the status of Maoris in New Zealand or native Hawaiians and Native Americans in the United States. Even Christianity seemed threatened, not to mention land rights and chiefly authority. A new Fijian movement, the Taukei (Landowners) movement, responded with threats and demonstrations against the cabinet and the Indian community; the consequent disorder then provided the justification for the mid-May military coup led by the 38-year-old Lieutenant Colonel Sitivini Rabuka, third in command of Fiji's army. Before the Bavadra government's intentions could be tested, it was dissolved and its members placed in detention. Suspending the constitution, Rabuka declared himself head of state.

In the next few months, it seemed possible that Governor-General Ganilau, who at first had condemned the takeover and refused to step aside, would be able to bring about some new accommodation—a successor regime in which Bavadra's forces would at least be represented and the task of drafting a new constitution would begin. In mid-September, however, in the midst of seemingly promising negotiations to this end, Rabuka, now a brigadier general, staged his second coup. Removing Ganilau from office, he assumed full charge. In short order, the Rabuka government suspended civil liberties, banned political activity, and, to ensure strict observance of the sabbath, prohibited Sunday sports, entertainments, film shows, and outdoor recreational gatherings.

In reaction to this new blow to constitutionalism and civil government, the Commonwealth excluded Fiji, Western criticism burgeoned, and tourism fell sharply. Capital flight and Indian emigration mounted, especially among professionals and the technically trained, while strike actions by Indian cane cutters, small shopkeepers, and others threatened to bring an already declining economy to a halt. In these disturbed and disturbing circumstances Rabuka retreated somewhat, and in December 1987 Ganilau and Mara agreed to take the helm. In what was now the Republic of Fiji, Ganilau became president, Mara, prime minister. Rabuka, now head of the armed forces, remained in the cabinet in charge of defense and security.

The Return to Constitutionalism

Over the next years as a new constitutional order was being devised, the conflict was less a struggle between Fijian and Indian views, which were largely ignored, than a tug-of-war between Rabuka on one side and Mara and Ganilau on the other. Like-minded in wanting absolute and unassailable protection for Fijian preeminence, they differed sharply on how to accomplish this. Rabuka, the commoner, the impulsive, outspoken, blunt, frequently inconsistent radical nationalist, reveled in his wide popular support. A Methodist elder, he believed his actions to be inspired by God, while his views of how authority should be exercised had been shaped by his active and successful military career. As one observer put it, his behavior looked "like the ambitious posture of the Melanesian Big Man rather than the deferential obedience a Fijian displays in the presence of his traditional chiefs."[2]

As the highest of high chiefs, Ganilau and Mara (the latter also widely respected as the region's senior statesman) had good reason for concern with Rabuka's threat to the paramountcy of the high chiefs and to habits of obedience. Conservatives to the core, they favored moving deliberately through the consultation and consensus process, earlier dubbed the "Pacific Way" by Mara himself. They were more sensitive than Rabuka to the dangers to the Fijian economy posed by international disapproval and Indian desperation. And they placed high value on preserving the forms of constitutional and civilian government, even while parliament had ceased to exist and the old constitution had not been replaced.

As the country moved slowly toward a new constitution, Ganilau and Mara, perhaps persuaded that Rabuka had outstayed his usefulness, moved sometimes with unusual directness to hem him in politically. Equally directly, Rabuka made it evident that he was determined both to remain in politics and to lead the country. Meanwhile Fijian divisions continued in other quarters. The western chiefs persisted in their claims against the eastern establishment and a bitter power struggle developed within the dominant Methodist Church over how far sabbath observance should be carried.

Despite political uncertainties, Fijians had some reason for satisfaction. In 1988 the economy picked up sharply and began to return to normal after a worrisome period of high inflation, a devalued currency, and reduced foreign investment. By the end of 1988 both the anticipated demographic trends and a substantial Indian emigration had restored the numerical edge the Fijians had lost in 1946; with an estimated 23,000 to 28,000 Indians having departed, official statistics for 1990 gave Fijians 48.9 percent of the population and Indians 46.2 percent.[3]

Meanwhile, protests and the occasional strike aside, the Indian community had remained generally passive. The FLP lost some of its Fijian character with Bavadra's death in November 1989, the subsequent defection of his widow and successor, Adi Kuini Bavadra, herself of high chiefly rank, and the accession to the leadership of Indian trade unionist Mahendra Chaudhry. NFP as well as FLP leaders had rejected invitations to participate directly in constitutional drafting panels; the NFP, however, made written submissions reiterating its traditional positions on noncommunal voting and parliamentary sovereignty.

The New Political Order

The wholly Fijian-drafted July 1990 constitution, even though more moderate than appeared likely three years earlier, is essentially unqualified in its protection of Fijian primacy. Subject to review in seven years, it requires both president and prime minister to be Fijians, with the president appointed for a five-year term by the Great Council of Chiefs. Twenty-six upper house seats are reserved for Fijian chiefs appointed by the president on the advice of the Great Council; the other nine are to be non-Fijians appointed by the president "on the basis of his own deliberate judgment." In the lower house 37 seats are reserved for Fijians, 27 for Indians, and 4 for General Electors; voters, casting one ballot, must choose only from among candidates of their ethnic community. Half of all public service jobs are reserved for Fijians, 40 percent for non-Fijians. The army is given "overall responsibility to ensure at all times the security, defense, and well-being of Fiji and its peoples." However, the final text dropped an earlier provision making the commander-in-chief automatically a member of the lower house and minister responsible for security. Also dropped was an unqualified assertion of Fiji's national identity as a Christian state. Instead, accompanying the preamble's affirmation that "the state accepts that Christianity plays a prominent role in the lives of indigenous Fijians" is the assurance that it also accepts "that other religious groups should be free to practice their own religion."[4]

Once the constitution was promulgated, political activity quickened in anticipation of the elections, originally scheduled for 1991 and finally held in May 1992. In this period, Rabuka's transformation from military commander to populist politician was completed. In October 1991, despite his continuing and frequently vocal differences with Mara and Ganilau, he won the leadership of the recently organized chiefly political party, the Soquosoqo Vakavulewa Ni Taukei (SVT). Significantly, the two candidates he defeated were both high chiefs: one of them the incumbent SVT president, Adi Lady Lala Mara; the other Ratu William Toganivalu.

Meanwhile, on the Indian side, the forces arrayed differed little from those of the precoup period. The Indian component of the Alliance had been replaced by a new party committed to political participation. The NFP and the FLP continued as the principal components of the opposition, each competing with the other but both highly critical of the new constitution as discriminatory and both pledged to boycott the elections. They abandoned this pledge, however, running full slates in the election, the FLP under Chaudhry, the NFP under its longtime leader, Jai Ram Reddy.

In the elections the NFP and the FLP divided the Indian seats almost equally between them, while the SVT won 30 of the 37 Fijian seats in the 70-member house. Although the SVT itself lacked the majority necessary to organize the government, an SVT-organized cabinet could probably expect support from the five-seat General Voters Party and a few independents. More significantly, however, the party was also divided by rivalry among three members of parliament for the prime ministerial nomination: Rabuka, Toganivalu, and Mara's protégé, former deputy prime minister Josefeta Kamikamica. Although Rabuka had the lead in the SVT and the support of the General Voters Party, to gain a majority he needed Indian support as well. A deal with the FLP got him the necessary votes. In exchange, Rabuka made a promise, subsequently broken, to institute an early review of the constitution and of recent decrees restricting trade union rights.

No such deals were necessary after the February 1994 elections. Rabuka by then had consolidated his control of the SVT, which took 31 of the Fijian seats; Kamikamica, who had organized a breakaway party, lost in his own constituency while his party—the Fijian Association—won only five seats. Toganivalu died suddenly in the course of the campaign. On the Indian side, accommodation seemed to be winning the day, with the NFP taking 20 seats against 7 for the more confrontational FLP. Underlining the weakness of Rabuka's Fijian opposition, Kamikamica was once again defeated in a by-election in October 1995. Rabuka himself has survived his involvement in an extramarital relationship; difficulties with his General Voters Party coalition partner; cabinet fissures over how to respond to the French nuclear tests; and the opposition aroused by his proposals to abrogate the sabbath observance law, remove the ban on gambling, and import large numbers of Chinese from Hong Kong.

For the time being, at least, the Indian political parties joined in the parliamentary opposition seem to have concluded that playing the game of politics by Fijian rules offers more than not playing it at all. Among Fijians, extremists, who favor such measures as expelling all Indians, have not disappeared. The mainstream, however, given the more absolute protection offered by the constitution, seems prepared to agree that Indians do have rights, even if not equal ones. Thus, the SVT constitution, while defining the party's primary objective as safeguarding Fijian interests, pledges to promote these interests "in association with other ethnic communities in Fiji."[5] The political chameleon Apisai Tora, in 1987 a leader of the extremist Taukei movement, in 1991 presented his new, largely Western-based All National Congress Party as a multiracial one, declaring: "The people of Fiji, whether we are Fijians, Indians or Europeans, or whatever, are here and here to stay. There is no place else to go."[6] Rabuka intermittently stresses the importance of Fijian-Indian cooperation although, in his usual erratic manner, veering between inducements, such as possible Indian participation in a national unity cabinet and threats of another coup should Fijians continue to feel economically threatened by Indians. Relations between the two communities remain uncertain and suspicions and apprehensions are easily aroused on both sides. Moreover, the problems dividing them will come increasingly to the fore with the approaching expiration of the land tenancy law in 1997 and as deliberations continue on the constitutional review required by 1997.

Meanwhile, whatever the developments in Fijian-Indian relations, changes in the Fijian sociopolitical order have been accelerated by Rabuka's challenge to the chiefly system. High chiefs continue to occupy the positions to which tradition assigns them. But there do not seem to be any on the horizon who, like Mara and others of his generation, combine inherited status with aspirations for political authority and talent in exercising it.

Papua New Guinea

Although the 1976 settlement had seemed to resolve the secessionist problem in North Solomons province, discontent persisted with the way Bougainville Copper Ltd. (BCL) conducted its massive mining operation at Panguna on the province's largest island. Provincial leaders, lay and clergy alike, charged that the central government and the BCL were receiving the lion's share of profits from the mining operations, which were placing heavy burdens on the local people through their impact on the environment. To grievances against the central government and the company were added disputes, almost as bitter, among titleholders. "First we were against the company," lamented a Bougainville member of parliament, "then the PNG government, then we were at each others throats: clan against clan, families against families, brothers against brothers."[7]

The Clash of Values

As described in a statement issued by the diocesan priests of Bougainville, the secessionist violence that flared up in 1988 was "the result of the imposition of Western concepts of landownership, value, and compensation upon a Melanesian social structure."[8] Basic to this clash of values between imported law and Melanesian practice was the allocation of subsurface resources to the state according to Australian practice, in defiance of Melanesian principles allocating them to the landowner. In practical application, this meant that the central government was the principal negotiator with BCL and set aside for itself a much larger share of the wealth produced at the Panguna mine than was allocated to the provincial government or to the affected landowners.

The equities involved in the distribution of returns from the Panguna operation were substantial. From its opening in 1972 until its closing in 1989, it was the largest industrial enterprise in Papua New Guinea. Its output constituted 44 percent of Papua New Guinea's total exports. Thirty-five percent of Panguna's profits went to the central government, its income from dividends, taxes, and royalties amounting to about 16 percent of its internally generated revenue between 1973 and 1989. The 5 percent allotted to the provincial government, although constituting one third of its revenues, seemed paltry by comparison. One landowner claimed that his province is like "a fat cow to be milked for the rest of the country." Even more insignificant by comparison were the returns in compensation, royalties, and public services to the landowners, amounting to a little more than one percent of the total profits from 1973 to 1989.[9]

Nor was the local population the prime beneficiary of increased employment opportunities. At the higher technical levels expatriates, mostly Australians, predominated. Of the roughly 3,000 PNG citizens employed in 1988, two-thirds were from other parts of the country.[10] Contemptuously named "redskins" by the very black Bougainvilleans, they brought with them all the problems associated with alien, disesteemed, predominantly single male migrant labor.

Added to grievances over the distribution of benefits was the impact on the environment of the huge, open-cut copper mine: "a polluted river, a mountain of tailings, and one of the largest holes ever made by man."[11] For a people long dependent on hunting and fishing as well as subsistence agriculture, the repercussions extended far beyond the mine site itself. Means of livelihood were being destroyed, links to the past were being broken, and even the future was being demolished. "Land is marriage—land is history—land is everything," a local landowner mourned. "If our land is ruined, our life is finished."[12]

The Persistence of Secessionism

Wherever mining operations have developed in Papua New Guinea, similar issues have been important. That the problem exploded when and where it did reflected the fact that large-scale mineral exploitation in Bougainville, having begun in 1972, was much further advanced there than elsewhere in PNG, and thus its impact on the environment and its profits were easier to see. That pressures for change moved so swiftly into secessionist channels reflected the special intensity of Bougainville localism, richly fed by the belief of the indigenous people in their unique identity, their superiority, and their conviction expressed in the words of former provincial premier, Joseph Kabui that "ever since God created the Universe, Bougainville has been separate, has been different."[13]

Distance from the Papua New Guinea mainland has encouraged the belief in Bougainville's separate identity. Geographically part of the Solomons archipelago, Bougainville was not transferred from German to British rule until the territorial exchanges of the late 1890s. Even after that, until late in the colonial era, such educational and administrative services as existed in this remote territory were provided not by British officials but by the Christian missions—dominantly Catholic—that came to Bougainville from the Solomons and maintained their principal ties there.

Incendiary tensions also arose from conflict over land titles and from dissatisfaction over the way landowner interests were being pursued by the Panguna Landowners Association (PLA), their official representatives. Opponents of the status quo charged that many were unjustly landless because traditional matrilineal descent patterns had been ignored in the original negotiations with Bougainville Copper Ltd. The PLA executive committee was accused of conceding too much to BCL, pursuing selfish interests, and misusing the income of a trust fund intended to be spent for community projects. Toward the end of 1988 the PLA leaders, now widely discredited, were voted out of office. The younger element, followers of the militant Francis Ona, a former BCL surveyor, took over the reins, constituting themselves the new PLA in a meeting attended by provincial premier

Kabua and national-level representative Father Momis. Embarking on a much more radical course, the new leadership's demand for an enormous sum in compensation (more than the central government had received since the mine opened) was generally dismissed as absurd. The attention attracted by this extravagant demand also obscured demands related to landowner grievances against one another, among them a proposal for a new survey of land titles.

The Resort to Violence

Their compensation demand unsuccessful, Ona and his followers in December 1988 responded with a sabotage campaign in the name of the Bougainville Revolutionary Army (BRA). By the end of 1989 BRA sabotage (employing explosives stolen from the mine) had cost the company two complete power shutdowns and considerable damage.

The government's initial response reflected the influence of accommodationists, like Momis in the national cabinet and John Bika (later the victim of a BRA assassin) in the provincial one, who supported demands for redistribution of mining returns but opposed secession and who sought to restore peace and retain Bougainville in the nation by concessions on both sides. Ona remained intransigent, however, and his forces grew, augmented by purely criminal *raskol* bands. As violence spread, the badly stretched police grew increasingly restive under a burden too heavy for them to carry without military assistance. Proponents of an all-out military response began to exert greater influence over government policy. In March 1989, for the first time, troops were sent in to augment the police. In June the BRA was outlawed and a state of emergency was declared. In December 1989 the mine was mothballed. Although intermittent negotiation efforts were fruitless, military action was no more successful, as the lawless and brutal behavior of the troops antagonized many who had formerly favored compromise. In May 1990, when Ona declared an independent North Solomons Republic with himself as interim president, he was able to enlist Joseph Kabui as premier and obtain the support of a significant segment of the local political and clerical leadership.

Three prime ministers—Rabbie Namalui, Paias Wingti, and Julius Chan—have struggled unsuccessfully to resolve the Bougainville problem. Armed force, military withdrawals, cease-fires, negotiations, and blockades have failed to put an end to violence or bring Ona and his lieutenants out of the bush. Under Chan seemingly much improved prospects for a peace conference encouraged the organization and deployment in Bougainville of a peacekeeping force to which Fiji, Tonga, and Vanuatu contributed and for which Australia and New Zealand provided logistic support. But this too came to naught. Nevertheless, by the beginning of 1993 the government had succeeded in establishing its control over Buka Island and substantial parts of Bougainville. Meanwhile, the BRA was weakened not only by government measures but also by its divisions and the war-weariness of the population. Early in 1995 a Bougainville Transitional Government was inaugurated, charged with bringing all of the provincial factions together and developing a new relationship with Port Moresby. Ona and his followers within the BRA have remained intransigent, but more moderate BRA members have been engaged in a dialogue with transitional government supporters, which, it is hoped, will lead to actual peace negotiations.

Repercussions at Home and Abroad

The protracted Bougainville problem has provided opposition politicians with ample opportunities to attack government competence. It has also contributed to already existing problems in the defense forces—indiscipline in the ranks to the point of strikes over pay and working conditions and tendencies at higher levels to act without government authorization. But it has not been an important source of government instability. Namalui's loss of office in 1992 after four years as prime minister was much more the result of corruption charges than of his government's failure to end the civil war; economic mismanagement and his unconstitutional behavior were preeminent among the issues that brought Wingti down in 1993.

The loss of Panguna's revenue has not been the disaster to the national economy that might have been expected. Instead, several years of high economic growth followed the mine's closing and Panguna's lost mineral production was more than replaced by the opening of rich new mines elsewhere, the expansion of existing ones, and oil discoveries. Although the expenditures imposed by the rebellion may have contributed to the government's serious budget difficulties of the last few years, they have not been a major factor.

Bougainville itself has borne the heaviest costs. Migrant laborers may have been the first sufferers from the loss of 3,000 jobs at the mine, but its closure also caused the collapse of the local service companies dependent upon it. A sharp decline in tree crop exports—an important component of the provincial economy—resulted from both the government blockade and the flight of plantation workers and managers. Villagers have been terrorized by both sides, their homes and gardens destroyed, while persistent disorder has provided cover for the depredations of lawless *raskol* bands. The report of an Australian parliamentary delegation, visiting in April 1994, painted a grim picture of destruction, dislocation, an acute shortage of health facilities, significantly reduced access to education, and drastic declines in smallholder cash crop production. Some amelioration has been provided by government "care centers" for people driven from their villages. For the 40,000 to 50,000 people who have been sheltered in such centers, however, life has been at best uncomfortable and debilitating.[14]

Although Papua New Guinea's domestic politics have been very little affected by the Bougainville crisis, this has not been true of its foreign relations. The Solomon Islands government was quick to declare that Papua New Guinea's difficulties were purely domestic and that it had no intention of becoming involved. Over time, however, abstract commitments tended to be outweighed by the accessibility of the Solomons to BRA elements seeking safe haven, supplies, and a base for publicizing their case; a desire by some in Honiara to cultivate the goodwill of a potentially independent North Solomons neighbor; and anger at PNG military incursions. With neither country capable of fully controlling the sea boundary, Papua New Guinea's determination to close off these opportunities led to seizures in Solomons waters and raids into Solomons territory in which houses and gardens were damaged or destroyed and, in one case in September 1992, two villagers were killed and a child injured.

Relations between the two countries, which began to deteriorate in 1989, had reached a new low in the last months of 1992 when the Solomons, to no great avail, appealed to both the United Nations and the South Pacific Forum to take up the case. Talks between the two governments, broken off after the September 1992 incident, were then resumed. Thereafter, elections and a change in government in the Solomons created a distinctly improved atmosphere, with the new prime minister, Billy Hilly, proving less confrontational than his predecessor, Solomon Mamaloni, playing an active part in efforts to bring about PNG/BRA talks and to establish a multilateral peacekeeping force. The improved atmosphere survived Mamaloni's return to office in October 1994.

Notes

1. Brij V. Lal, *Broken Waves: A History of the Fiji Islands in the Twentieth Century* (Honolulu: University of Hawaii Press, 1992), p. 222.

2. John Garrett, "Uncertain Sequel: The Social and Religious Scene in Fiji since the Coups," *Contemporary Pacific*, Spring 1990, p. 88.

3. FBIS-EAS, December 14, 1989.

4. Ralph R. Premdas, "Fiji under a New Political Order," *Asian Survey*, June 1991, p. 557.

5. "Melanesian Review," *Contemporary Pacific*, Vol. 4, no. 2 (1991), p. 387.

6. Ibid.

7. *Pacific Islands Monthly*, November 1992, p. 17.

8. Peter Polomka, ed., *Bougainville: Perspectives on a Crisis* (Canberra: Australian National University, 1990), p. 95.

9. Ibid, pp. 43–45; *Pacific Islands Monthly*, January 1989, p. 15.

10. *Far Eastern Economic Review*, August 3, 1989, p. 30.

11. George K. Tanham and Eleanor S. Weinstein, *Papua New Guinea Today: June 1990* (Santa Monica: RAND Corporation, 1990), p. 13.

12. R.J. May, "Papua New Guinea's Bougainville Crisis," *Pacific Review*, Vol. 3, no. 2 (1990), p. 174.

13. Matthew Spriggs, "Alternative Prehistories for Bougainville," *Contemporary Pacific*, Fall 1992, p. 269.

14. *Report of the Visit of the Australian Parliamentary Delegation to Bougainville, 18–22 April 1994* (Canberra: Australian Government Publishing Service, 1994).

PART FOUR

Coping with the Outside World

8

Constraints, Assets, and Initiatives

Their minuscule size, their exceptionally narrow resource base, their remoteness from centers of global commerce or other international preoccupations, their inability to maintain or attract significant diplomatic establishments—all these factors have combined to push the island states close to the bottom of the global priority list. To some extent, moreover, international inattention to the island states has been a tribute to their virtues—the absence of the wars, revolutions, and pogroms that have attracted greater outside attention to other countries. External concern with Fiji's coup was as exceptional as the event itself, and it quickly evaporated.

In their relations with other countries, however, the island polities have not been without assets. The linkage they make between their strategic interests and the stability and well-being of their island neighbors encourages Australia and New Zealand to provide assistance and support. The global competition for influence that spread to the South Pacific in the mid-1970s enhanced island bargaining power. In the same period a more enduring asset—their new jurisdiction over Exclusive Economic Zones (EEZs) extending 200 miles from their shores—gave the island states new leverage in the intensifying competition for access to diminishing global fish stocks. And, magnifying such assets as they have, the success of their initiative in joining with Australia and New Zealand to create a regional organization, the South Pacific Forum, has provided the 14 independent island states with a stronger voice internationally than they could possibly claim individually.

Constraints

As economic partners the Pacific island states have little to offer other countries, and expectations of economic gain have played little part in the relations external states maintain with them. The region's rich stock of tuna is one exception. New Caledonia's nickel is another; its reserves, constituting an estimated 40 percent of the world's total, are often cited as influencing French opposition to independence. Overall, however, a much larger share of the world's resources flows into the Pacific island states in the form of aid than flows out in the form of exports or return on investments. Except for nickel, none of the important island mineral exports—copper, gold, phosphate—contributes an essential or even significant share of world consumption.[1] Similarly, although copra and other coconut products dominate exports in a number of island countries and are at least significant in almost all, the region as a whole provides only 6 percent of total world production. As customers, the island countries may be significant to particular firms or producers. But they are not more generally important. For example, although most island imports come from Australia and New Zealand, this does not give the islands high rank as customers there. In the 12-month period ending in June 1992, they took only 3.67 percent of New Zealand's exports.

Private foreign capital in significant amounts has been attracted only to Papua New Guinea. In the other states, local markets are small and foreign markets are hard to reach. Even in Papua New Guinea, investors attracted by the country's mineral wealth must also weigh the risks posed by lawlessness and the difficulties of negotiating agreements that traditional landowners will continue to accept as binding over time. Nor do foreign investors bear all the risks. All too often, unscrupulous foreign entrepreneurs, attracted to one or another island country by prospects for short-term profits, have engaged in practices that have been not only exploitative but also environmentally destructive and that have contributed to official corruption.

Foreign aid is thus a central element in the economic life of the island states and in their external relations. Although absolute amounts are not large by world standards, on a per capita basis Pacific island polities together receive more aid than any other world region. One specialist in island economies estimates that aid typically amounts to 40 percent of government revenue and between 35 percent and 50 percent of government expenditure.[2] The larger share is provided by former and present metropoles and Japan, but other countries and a variety of international organizations and nongovernmental groups are also among the donors. Aid is supplemented by preferential trading relations with Australia and New Zealand and with Western Europe, France especially. Remittances from nationals overseas—predominantly in New Zealand and the United States—make another considerable outside contribution to island revenues.[3]

Island dependence on foreign aid is likely to continue for some time. Specialists in island economies are pessimistic about the prospects for

self-sustaining growth: expanding population is putting additional pressures on limited resources, and modernization has long since excluded
reversion to the subsistence economies of the past. Only Papua New
Guinea, Fiji, and perhaps the Solomon Islands are seen as likely, eventually,
to achieve self-sustaining economies at tolerable levels of living.

The paramount role of donor countries in decisions about amounts and
kinds of aid raises nettlesome issues for island countries. Although aid has
been among the factors encouraging island identification with the West,
donors have not found aid suspension a particularly efficacious means of
influencing recipient states' political policies. French use of aid reduction to
put pressure on an unfriendly Vanuatu in 1981 and 1987 further embittered
the relations between the Lini government and Paris. Australian and New
Zealand suspension of aid in response to the Fiji coup had no impact on
political developments in Suva, while stimulating another round of island
criticism of what was seen as the domineering tendencies of the two states.
Economic policy, on the other hand, is more readily shaped by the conditions on which aid is granted. Today, however, the principal dangers of aid
dependency lie in the budgetary pressures on donor countries to reduce
external expenditures, a decision the island polities are in no position to
influence.

In the global arena, the island states see themselves as all too frequently
overlooked. Thus although they regard themselves as quintessentially of
the Pacific with a proprietary interest in its waters, their views in this regard
have rarely influenced the thinking of proponents of Pacific-wide organization. It was not until 1992 that a Pacific Forum representative, Secretary-
General Ieremia Tabai of Kiribati, was invited to address the ministerial
meeting of the intergovernmental Asia-Pacific Economic Cooperation
(APEC) forum, organized in 1989, and not until 1993 that a Pacific island
state—Papua New Guinea—was invited to join. "It is as though the Pacific
Ocean was nothing but a vast empty space, inconveniently getting in the
way of attempts to forge closer ties among rim nations," complained Tabai.[4]
Seeing indifference coupled with exploitation, Vanuatu's Prime Minister
Carlot has argued that even while major powers preoccupied with the
Pacific rim ignore what happens inside, "a growing number of governments
or large private companies are pursuing our smaller nations either to seek
their votes in international organizations or to come and exploit the few
resources they may have."[5]

Assets

Complain as they may of neglect, the island states have not had to go
it alone. Former metropoles have continued to accept some responsibility

for former colonies; in other cases continued dependent status has ensured the attention and support of administering powers and given them some interest in the surrounding independent island world.

Australia and New Zealand have attached particular importance to helpful relations with their former colonies as well as to participation with the island states in regional affairs. From their earliest days as political communities, they have regarded the stability and security of the islands and the goodwill of their rulers as important to themselves. Their activism has ranged from participation in imperial expansion to support for the independent states. But it has never been abandoned. The islands in consequence benefit from assistance that goes beyond economic aid to support for maritime surveillance, intelligence exchange, military training, and joint military exercises and extends to the willingness of both countries to use their middle-power rank and more extensive international diplomatic presence to support island interests.

During the cold war, the compulsion to overlook no sparrow, however small, spurred competition for island favor, not only West versus East, but also Moscow versus Beijing, and Beijing versus Taipei. For Australia and New Zealand proximity was important in the effort to maintain the South Pacific as a "Western lake," while the United States and France also valued the contributions of their respective territories to their larger security structures: for the United States, arrangements for basing and testing and the denial of military access to others; for France, the contribution made by testing in Polynesia to maintaining the *force de frappe*.

This interest was heightened in the 1970s and 1980s when Western alarm over more active Soviet interest in the region gave the islands new bargaining power. In 1989 New Zealand diplomat Denis McLean, commenting on island dexterity, pointed out:

> As early as 1976, Tonga swiftly turned Australian and New Zealand concern about the prospect of Soviet assistance with the development of the airfield at Nuku'alofa into a substantially increased aid package in return for rebuffing the Soviets. Neither Kiribati nor Vanuatu appears to have done themselves harm from recent careful negotiations with the Soviet Union over fisheries; they have gained substantial revenue and are now taken more seriously by others on the Pacific stage.[6]

In the same period, the importance of the Pacific island states to other governments was also boosted by the alterations in the international law of the sea that gave coastal states exclusive jurisdiction over all living and mineral resources in a zone extending 200 nautical miles from their shores.[7] For the first time, the island states had a tangible bargaining asset—their Exclusive Economic Zones (EEZs), totaling 6 million square miles of the

Pacific. In this vast area, they could impose licensing fees and other require-ments on distant-water fishers, hitherto free to operate on their own terms in exploiting the region's rich stocks of tuna.

Japan, South Korea, and Taiwan, whose nationals dominated among distant-water fishers in the South Pacific, now found it necessary to deal on these matters with Pacific island governments and useful to offer aid projects related to their fishery interests. Japan in the following years became an aid donor second only to Australia. Much of the Soviet Union's new activity in the region also reflected interest in access to its tuna re-sources.

Initiatives

The creation of the South Pacific Forum in 1971 represented early recognition by the then-independent island states—Cook Islands, Fiji, Nauru, Tonga, and Western Samoa—of two imperatives. One was the need to join together if they were to make their voices heard; as Ratu Mara, one of the Forum's founders, observed, "When we were considering our con-stitutional evolution from our colonial status through to self-government and eventual independence, we realized that, with our small size and relative isolation, independence could not be entirely viable without some sort of association with our neighboring Pacific island territories."[8] The other was the desirability of responding to the interest of Australia and New Zealand in being accepted as part of a South Pacific community, given the weight their membership would add to the nascent organization.

The South Pacific Commission (SPC) established in 1947 provided both a precedent and a point of departure. Although in 1971 the island states had not achieved the equal participation in the SPC that they enjoy today, their early limited role still gave their leaders experience in operating in the multilateral context and established links among them that, in Southeast Asia for example, took much longer to develop. But in the eyes of island leaders, the SPC retained its colonial taint, while the exclusion of political issues from its agenda left them without a recognized collective voice on such matters as French nuclear testing. It was this voice that the Forum provided.

Their many shared characteristics facilitated cooperation among the Forum island states. The Francophones of Vanuatu excepted, English is the common tongue of their elites; representative democracy, mostly based on the Westminster model, provides the constitutional framework for national politics; Christianity is universal and churches are strong, active, and influ-ential. Common allegiance to the "Pacific Way," which emphasizes a shared tradition of respect for the views of others, harmony, and decision by consensus, also links the islands, even though, as with Christianity, its precepts are not always observed in practice.

In the interests of regional comity, the Forum sedulously avoids interference in the domestic affairs of its members. It broke this rule only once, in the case of Fiji, with results that seemed to demonstrate its worth. Meeting on schedule in 1987, two weeks after the Fiji coup, the Forum (which Fiji did not attend) expressed "deep anguish" over the "overthrow of the elected government" and offered to send a delegation headed by Australian Prime Minister Bob Hawke "to assist the governor-general in his efforts toward a peaceful and satisfactory solution."[9] Within the Forum, particularly among the Melanesians, however, there was distaste for this intervention in Fijian domestic politics. PNG Prime Minister Wingti sharply characterized subsequent Fijian rejection as an embarrassment that could have been avoided had more weight been given to Melanesian opinion. The mistake was not repeated: Fiji was not on the 1988 Forum agenda, Ratu Mara once again led his country's delegation, and Bavadra's representative was generally shunned.

Perhaps with the lesson of Fiji in mind, the Forum at its 1992 meeting rejected Honiara's appeal to put Bougainville on the agenda, accepting PNG's insistence on defining the problem as purely internal. Two years later the Bougainville peacekeeping force was organized independently of the Forum. That PNG's own Sir Julius Chan had sought its establishment removed what would otherwise have been an insuperable obstacle. That his request was so readily granted demonstrated the strength of the regional solidarity the Forum has fostered.

South Pacific geography has facilitated regional cooperation, even though it limits the fields in which it can be effective. To be sure, small aid-dependent countries separated by vast distances and poorly served by air and sea transportation do not have much to offer one another by way of economic cooperation or common defense. However, the distances that inhibit contact and communication in most of the region have also precluded the border disputes that in other parts of the world have led to generations, even centuries, of conflict. An exception is the intermittently troubled relationship between Papua New Guinea and the Solomon Islands stemming from Bougainville-Solomons proximity and ethnic ties.

There are other although less troublesome fissures in the region. National interests vary and sometimes conflict; tiny states have different problems from larger ones. Melanesians resent Fijian and Polynesian pretensions to superiority, encouraged by the colonial rulers, and their prominence in regional structures, gained through earlier independence. Polynesians in turn resent Melanesian claims to importance based on larger size and more ample resources. When Papua New Guinea, Vanuatu, and the Solomon Islands joined together in the Melanesian Spearhead to support the Kanak cause, it seemed possible that this would be the first step toward a Melanesian/Polynesian breach. Australia and New Zealand are not always fully accepted partners; on occasion both have been criticized as overbearing and, Australia in particular, as overly anxious to bring Forum positions into conformity with those of the United States.

So far in the Forum's history, however, cohesive factors have outweighed divisive ones. It has taken in stride the emergence of new leaders and a changing array of problems. What seemed to pose the most serious threat to Forum cohesion, the Melanesian/Polynesian divide, has not fulfilled pessimistic expectations. Competitively generated efforts to organize a Polynesian counterpart to the Melanesian Spearhead never got off the ground. Spearhead unity has been intermittently undermined by PNG/Solomons differences, nothing has replaced its pre-Matignon Kanak-centered energizing mission, and Fiji has continued to reject membership, unwilling as its leaders are to identify their country fully with Melanesia.

The Forum's membership has grown steadily in parallel with the island state's accession to independence. As of the spring of 1996, it has 16 members whose leaders meet at an annual summit to discuss issues of regionwide concern and to arrive at regional positions. The Forum commands the support of a number of technical bodies including a secretariat (formerly the South Pacific Bureau for Economic Cooperation, or SPEC) active in matters of common economic concern, the Forum Fisheries Agency (FFA), and the South Pacific Regional Environment Program (SPREP), a cooperative effort with the SPC.

As it has grown in numbers, expanded in functions, and demonstrated its survivability, the Forum has also developed a more substantial international personality. It has been accorded observer status at the UN General Assembly; it has joined forces with similarly situated countries in the Alliance of Small Island States; its secretariat has been engaged with its Association of Southeast Asian Nations (ASEAN) counterpart in a number of cooperative ventures; and since 1989 in a variation of ASEAN practice, Forum summits have been followed by meetings with officials representing principal foreign partners.

The Forum, however, has not superseded the SPC. The latter, in the almost five decades of its existence, and within the constraints of a very small budget, has provided training courses and expert advice on agriculture, health, education, and similar subjects and served as a clearinghouse for the exchange of information. Even so, the expanding Forum role and the proliferation of institutions attached to it in one way or another have combined with the SPC's colonialist past to prompt arguments that duplication and waste could be more easily controlled and the colonialist taint removed by a single organization concerned with political as well as economic questions. The organization's defenders argue that membership in the SPC stimulates the interest of the former and present metropolitan powers in the region as a whole and gives an otherwise lacking regional voice to the remaining dependent territories while also strengthening that of the small Polynesian states. Stemming from a 1987 Forum-mandated study, a new mechanism was established in 1988, the South Pacific Coordinating Committee (SPCC), leaving otherwise untouched the institutions of regional cooperation.

The fate of the SPC remains moot. Expectations of its survival, however, have been strong enough to encourage France to invest in a new headquarters building in Noumea and the European Union to contribute $6 million to a five-year project to promote sustainable management of island fisheries. Britain, on the other hand, has withdrawn, although it will make a small contribution to the budget on behalf of Pitcairn Island. Eventually, the SPC rather than fading away may be transformed; proposals for amending its charter to permit the membership of such states as Canada, Chile, Japan, and South Korea are under consideration.

Notes

1. Although Papua New Guinea ranks ninth among the world's gold producers, in 1994 out of a global ouput of 1,899 metric tons, it produced only 61.5 metric tons.

2. Te'o I.J. Fairbairn, Charles E. Morrison, Richard W. Baker, Sheree A. Groves, *The Pacific Islands: Politics, Economics and International Relations* (Honolulu: East-West Center, 1991), pp. 48–50.

3. According to recent census figures, there are 14,400 Niueans in New Zealand.

4. *Pacific Islands Monthly*, November 1992, p. 9.

5. Ibid., December 1994, p. 11.

6. In a paper presented at a U.S. National Defense University conference, Washington, D.C., 1989.

7. Although it was not until November 1994 that the United Nations Convention on the Law of the Sea gained the 60 accessions necessary to bring it into force, its provisions respecting Exclusive Economic Zones had long since achieved acceptance as customary international law.

8. Quoted in a paper by Sandra Tarte presented at a conference in Apia, April 1989.

9. FBIS-EAS, June 1, 1987.

9

International Issues of Island Concern

In concerning themselves with international issues, the South Pacific Forum states have focused their pressure on the activities of outside powers deemed threatening to the island world. These perceived threats, while including lingering colonialism, have been largely related to the environment and resources—nuclear-weapons testing, waste disposal in the Pacific, and unregulated exploitation of fishery and forest resources.

Colonialism

As former colonies in a region where the sovereignty of external powers has not yet disappeared, the island states have perforce concerned themselves with colonial questions. However, official positions have been circumspect. Only Vanuatu under Lini took much interest in colonial issues outside the region. Within the region, official positions seemed to be required only in the rare cases where the decolonization process was not proceeding reasonably smoothly.

In island political circles, Indonesian sovereignty over Irian Jaya is sometimes attacked as imperialist. But governments in Port Moresby, accepting Jakarta's sovereignty as an accomplished fact, have made it quite clear to their neighbors that their differences with Indonesia are strictly bilateral. The United States has also been largely immune to serious anti-colonial pressures. During the Micronesian negotiations, aspects of U.S. policy were frequently criticized in the islands, but their outcome has been fully accepted. The absence of any serious demand for independence in the U.S. flag territories and their role in encouraging continued U.S. interest in the region are recognized.

In 1980 the secessionism that briefly threatened Vanuatu's progress toward independence became a common concern. Condemning "the illegal actions of a few," the Forum called upon France and Britain to fulfill their responsibilities for ensuring a smooth transition to independence on the agreed date. Of the island countries, however, only Papua New Guinea responded to Lini's appeal for assistance and sent troops to help quell the uprising. With this domestically somewhat controversial unilateral action in the background, Julius Chan, then-PNG prime minister, made an unsuccessful proposal in 1986 for the establishment of a regional peacekeeping force.

The Kanak struggle in New Caledonia has been a more protracted South Pacific concern. Kanak *indépendentistes* began to call for help from their neighbors early on, but their appeals began to have more resonance as Melanesian states achieved their independence. More militant supporters of their fellow Melanesians than their Polynesian neighbors, Papua New Guinea, the Solomon Islands, and Vanuatu joined together, first informally and then in 1978 as the Melanesian Spearhead, to support the Kanak cause. In the Forum, the Spearhead pressed for giving FLNKS observer status and for calling upon the General Assembly to reinscribe New Caledonia on the UN list of dependent territories. Other Forum members, however, were unwilling to go so far. Australia and New Zealand, as members of the Western alliance, were reluctant to challenge French policy so directly, while Fiji and some of the other island countries were restrained by the importance of France to their economies. Accordingly, Forum discussions of New Caledonia were reflected in rather bland communiqués. Typically affirming support "for the transition of New Caledonia to independence in accordance with the wishes of its people"[1] and urging steps in this direction, they avoided both strong criticism of France and proposals intended to bring French policy under UN review.

By 1986, however, Melanesian Spearhead arguments had become more persuasive as the conservative Chirac government seemed to be reversing the gains made under its Socialist predecessor. The Forum communiqué was correspondingly harsh. The new French government, it charged, "did not adequately recognize the aspirations of the Kanak people"; it "appeared committed to New Caledonia remaining a territory of France"; and its new policy "was a significant step backwards."[2] Although the Forum did not accord observer status to the FLNKS, it did agree to call for reinscription by the General Assembly, where it was supported by a large majority.

Two years later, the change in French policy signaled by the Matignon Accord evoked a positive response from the islands, the Forum communiqué indicating their continuing interest and willingness "to contribute positively to the process of reconciliation and cooperative endeavor now under way in New Caledonia."[3]

Atmospheric Issues

Environmental concerns are particularly high on the agenda of Pacific islanders. As Cook Island Prime Minister Geoffrey Henry has emphasized, "The subject of the environment is so important to us.— We in the Pacific have just the clear air that we breathe, the clear waters that we swim and fish in. That is all that God in His wisdom chose to give the peoples of this part of the world."[4]

Nuclear testing has been a particularly emotive issue for island governments and for an anti-nuclear movement widely supported by the churches, political leaders, trade unions, and other organizations. History played no small role in persuading Pacific islanders that their remote location and lack of power had made them the victims of great-power rivalries in which they played no part. Prolonged and well-publicized U.S. difficulties in dealing with the contamination and radiation sickness associated with the 1954 Bikini test and the hardships of evacuated islanders provided the background for the intense resentment that greeted the initiation of French atmospheric testing in 1966 in Mururoa, an atoll in French Polynesia, and that was not allayed by the shift to underground testing in 1975.

Against this background, proposals originated by Labor governments in Australia and New Zealand to declare the South Pacific a nuclear-free zone (SPNFZ) were warmly welcomed by island governments. The resulting 1985 Treaty of Rarotonga bound Forum signatories not to develop, produce, or use nuclear weapons and to bar nuclear-weapons testing or storage and nuclear waste disposal in their territories. The signatories were free to determine their own policies with respect to port calls by nuclear-powered or nuclear-weapons-capable naval vessels.

Three protocols were open to signature by the nuclear powers. In the first, France, Britain, and the United States were asked to apply the treaty's provisions in their Pacific territories; in the other two, the five nuclear powers were asked to pledge not to use or threaten to use nuclear weapons against signatory island members and not to test in the area.

The treaty had been drafted with careful attention to U.S. policy with respect to freedom of the seas and U.S. refusal to confirm or deny the presence of nuclear weapons on visiting warships. Nevertheless, Washington joined London and Paris in refusing to sign the protocols. Assurances that U.S. and British policies in the South Pacific would remain consistent with the provisions of the treaty and its protocols did little to allay Forum disappointment. Once more it appeared that island interests were being subordinated to great-power relationships.

When France temporarily suspended Pacific nuclear testing in 1992, its decision reflected post-cold war considerations, not island pressures. Greeting the decision warmly, the Forum continued to urge permanent suspension, which, it declared, would "contribute significantly to improving further the relations between France and the countries of the Pacific."[5] The announcement in May 1995 that France would soon embark on a new test series in Mururoa was thus seen as a deep affront.

However, while it was easy for the island states to join Australia, New Zealand, and much of the rest of the world in deploring French plans, it was more difficult to decide what action, if any, should accompany the rhetoric. The French government seemed unlikely to be swayed from its decision to test by any action taken by the island governments, individually or through the Forum. Moreover, France had added a new element to the situation by its commitments to end testing permanently after the contemplated six to eight tests had been completed and thereafter to adhere to the Comprehensive Test Ban Treaty and to sign the Rarotonga protocols. Since, if French commitments were to be taken seriously, Forum objectives with respect to nuclear testing in the Pacific would soon be achieved, how far should its members go in the interim in disturbing their relations with France, relations that France had been cultivating since the late 1980s by aid programs and heightened support for regional projects?

Even in the heat of their rhetoric, the responses of the island governments to the French challenge varied; few went beyond rhetoric to action. At one extreme, Vanuatu's prime minister Carlot argued that the tests fell within French sovereign rights and were an issue, if at all, only between Paris and Papeete. (He was subsequently rewarded for his supportive behavior by membership in the Legion of Honor.) At the other extreme, Nauru's president Bernard Dowiyogo demanded French expulsion from the Pacific while Tuvalu's deputy prime minister accused France of "the malicious and systematic destruction of the pillars of our very inheritance."[6] After the first French test on September 5, Nauru, Tuvalu, and Kiribati suspended diplomatic relations with France; earlier Niue and Western Samoa had joined them in withdrawing from the South Pacific Games held in Tahiti in August. Between the extremes, the Melanesian Spearhead was notably conciliatory. In a statement issued after its mid-August summit meeting, the Spearhead called upon France to reduce its tests to a minimum and commit itself to providing compensation for any resulting damage, while it emphasized the desirability of continued dialogue with France.

Forum members also found it difficult to agree on how to deal with the French dialogue relationship. Some were in favor of terminating the relationship. Others argued that maintaining it provided an opportunity to convey to France the depth of the region's anger. At its 1995 summit in mid-September, more than a month after the first test, the Forum expressed its outrage but left the dialogue relationship unchanged, warning only that any further tests would bring reconsideration. After the second test in October the relationship was suspended. However, as France declared its sixth test in January 1996 to have been its last and also reiterated its treaty commitments, it would not be surprising to see the suspension lifted in time for the 1996 summit.

Although nuclear testing in the Pacific has seemingly disappeared as an issue, the transit and disposal of nuclear and other hazardous waste remains a concern of island governments, convinced as they are that any harmful substance introduced in Pacific waters anywhere can arrive in due course on their shores. Criticizing Japanese waste disposal plans in 1980, Guam's first elected governor, Carlos Camacho, declared:

> For we who inhabit these islands, the sea is our farmland, our rangeland, and our forest.... We must take a united stand against any activity that threatens our rights to a clean ocean environment. Our people reap no benefit from nuclear energy, yet we are expected to share in its hazards.[7]

Reacting to both Japanese and U.S. plans, the 1984 Forum communiqué characterized the dumping of nuclear waste in the Pacific as intolerable and made clear the regional commitment to the principles of an environmental treaty being drafted by SPREP. Signed in 1986 by Western countries as well as the Forum states, the Convention for the Protection and Development of the Natural Resources of the Pacific prohibits the dumping of low-level radioactive waste. It is silent, however, on other potentially noxious materials that remain a subject of island concern.

In May 1990 the U.S.-owned Johnson atoll became a center of controversy when the United States announced a plan to destroy on this tiny uninhabited island not only the chemical weapons it had already stored there but also a smaller number stored in Germany. Later in the year, at his summit meeting with island leaders, President George Bush assured them that no environmental damage would result and invited the Forum to send a technical team to the island. In mid-1992 the team concluded that the plan posed no significant regional environmental threat.

The transit of nuclear and other hazardous materials on the high seas also stimulates fears of environmental damage. Tokyo's 1992 decision to ship a ton of plutonium from France to Japan and the security-dictated secrecy that veiled the route of the lightly guarded *Akatsuki Maru* evoked heated protests not only from Forum states but also from other nations on one or another of the ship's potential routes.

Another environmental concern has been global warming and the greenhouse effect. Here, even more than on other such issues, Pacific islanders feel a particular vulnerability to the actions of countries distant from their shores. On the one hand, rising sea levels caused by pollutants in the atmosphere may someday substantially reduce the land areas of some small island countries and completely inundate others. On the other hand, only the major industrial powers—the source of the carbon dioxide and other pollutants believed to be responsible for global warming—can undertake the expensive and complex actions that might prevent this outcome.

Resource Issues

Issues relating to the exploitation of island resources have involved both environmental and bread-and-butter considerations. For several decades, fishery matters, especially those relating to tuna, were in the forefront, the subject of both bilateral negotiations with foreign fishing states and regional action through the Forum Fisheries Agency. More recently there has been growing concern over the exploitation of forest resources.

Fish

In 1992 a million tons of tuna were taken from South Pacific waters, a harvest worth $1.5 billion providing 55 percent of the world's canned tuna. Obtaining a satisfactory share of the profit remains, from the island point of view, an objective far from achievement. Fishing fleets are mostly owned by the distant-water fishers of Japan, the Republic of Korea, Taiwan, and the United States. Foreign interests also dominate canneries and distribution networks; only a small percentage of the resulting revenue returns to the islands. Even so the ability of island governments to extract better terms has improved considerably since the establishment of EEZs gave them the authority to do so.

In seeking to profit from this new authority the island governments faced many problems. Island countries could not expect to replace the costly distant-water tuna fleets with their own vessels for many years, if ever. Regulation, licensing, and fees could contribute to the survivability of the resource and gain a share in the profits that distant-water fishers had long monopolized. But such regulation faced formidable obstacles. Little data were available to permit informed decisions about sustainable exploitation. Little was also known about the extent of the tuna resource, its migratory patterns, and even the size of the harvest. It was not clear that the interests of the island states coincided sufficiently to make a regional approach feasible. But with successful tuna fishing requiring rapid movement over long distances, how, except by regional arrangements, could island states cope with the fact that in six cases their EEZs overlapped and that, in others, the EEZ of one state could only be entered from the EEZ of another? Would those states with waters known to be especially tuna-rich, notably Papua New Guinea and the Solomon Islands, be prepared to share their returns with the less well endowed? Whatever regime the island states might adopt, would they be able to impose their regulations? Lacking surveillance and enforcement capabilities, could they count instead on the cooperation of distant-water fishers and their governments?

For the United States the answer to the last question was clearly negative. Uniquely among fishing nations, the United States rejected the authority of coastal states over highly migratory species and was required by its own law both to compensate Americans whose vessels had been seized for violating local regulations and to embargo fish imports from states making such seizures.

Briefly the solution seemed to lie in establishing a regional organization that included the United States whose membership would permit it to abide by regulations it could not otherwise accept. However, important island leaders continued to see the U.S. stance on unilateral regulation as an unacceptable affront. They feared U.S. domination and suspected that the U.S. position would mirror the interests of the American Tuna Boat Association (ATA), regarded by the islanders as unabashedly piratical. Accordingly, when the Forum Fisheries Agency was established in 1978, membership was confined to Forum countries. Moreover, with enthusiasm for regional regulation and exploitation largely confined to the resource-poor Polynesian countries, FFA functions were limited to collecting and analyzing relevant data, assisting in negotiations with distant-water fishers, and advising on conservation measures. Within a decade, however, reflecting changes on both sides of the Pacific, the United States was included in a regional fishing regime, and the FFA was given substantial enforcement and administering responsibilities.

The island states themselves had ventured further into cooperative relations. The states in whose waters tuna was most plentiful had agreed to establish uniform rules regulating observers on the vessels of distant-water fishers, log books, reporting requirements, and standardized identification.[8] An FFA-sponsored regional register of distant-water fishing vessels was showing encouraging results in terms of both the number of vessels registered and their compliance with the rules.

Meanwhile, developments in the U.S. fishing industry were undermining the foundations of the U.S. juridical position. In 1978, the United States was a minor participant in the South Pacific tuna fishery; by the mid-1980s, as eastern Pacific stocks were declining, the United States had become a major player with 60 American boats—each representing a huge investment—recording a tuna catch of over 100,000 tons in 1984. At the same time, on the U.S. Atlantic coast, the increasingly important bluefin tuna fishing industry was pressing for changes in U.S. policy, making it possible to exclude foreign tuna fishers, Japanese in particular, from the U.S. EEZ. On the West Coast similar pressures were coming from U.S. salmon fishers.

Cold war considerations were underlining the political costs of friction with the island states over unlicensed fishing by a growing U.S. fleet. In the mid-1980s, U.S. apprehensions were sharpened as the USSR, with its recently constructed tuna fleet, negotiated its first fishing treaties with island states—Kiribati in 1985 and Vanuatu in 1987. The unusually high price Moscow was prepared to pay suggested interests going well beyond fish. Although in the end neither treaty was renewed, the level of alarm was high enough to persuade the U.S. Congress of the importance of coming to terms with the islands.

The resulting South Pacific Fisheries Treaty of 1988 exchanged regulated access by U.S. tuna boats to island EEZs for licensing fees and technical assistance amounting to $12 million annually. Renewed at the expiration of its original five-year term, the treaty has been eminently satisfactory to both sides. Island pressures on the other distant-water fishers to substitute similar regional agreements for the present bilateral ones have, thus far, met with no success.

In contrast to their earlier differences over tuna, the island states and the United States have been allied against the threat to aquatic resources posed by gill- or drift-net fishing. First employed in the north Pacific and introduced in the south by Taiwanese and Japanese tuna boats, the nearly invisible nylon nets, suspended to a depth of 15 meters below the surface and often as much as 45 kilometers wide and 60 kilometers long, came to be known as "walls of death." Capturing not only the tuna or, farther north, the squid that were the intended prey but other varieties of fish, marine mammals, turtles, and sea birds as well, the growing gill-net fleet threatened to cause Pacific-wide extinction of the affected species.

FFA negotiations with Taiwan and Japan proved of no avail, the latter with its large aid programs and importance to island economies having distinctly the upper hand. Internationalization of the issue was a logical step; northern Pacific coastal states were also deeply concerned, and the U.S. Congress was already considering a ban on imports of fish caught in gill nets. In July 1989 the Forum, having already banned gill-net fishing in its own waters, called for banning gill-net fishing everywhere. This call was subsequently endorsed by the South Pacific Commission, the Commonwealth heads of government, and, at the end of the year, by a General Assembly resolution introduced by the United States. In a later action, the General Assembly called for the reduction in 1990 of gill-net fishing in the South Pacific, its termination there by July 1991, and its cessation everywhere by July 1992. In due course, although somewhat reluctantly, Japan and Taiwan accepted the global verdict, and gill-net fishing has ended around the world.

Timber

Although the islands' tropical timber, like their tuna, is unevenly distributed, as with tuna its commercial exploitation has become a common concern. The industry's recent growth in the South Pacific has resulted directly from problems elsewhere as rain forests have been denuded and log exports banned. Rising prices having compensated for difficult terrain and land-title complications, logging has become an important source of revenue for Papua New Guinea, the Solomon Islands, and Vanuatu. The foreign enterprises involved—mostly Malaysian—have brought many problems with them. Customary titleholders have been seduced by cash payments to accept deforestation of their land, and cutting has far exceeded sustainable yield. Continued logging at recent rates could exhaust Papua New Guinea's forests in 20 years, the Solomon Islands' in 8. Corruption flourishes in the award of licenses, collection of taxes, and enforcement of regulations. Moreover, important as it has been to their total revenues, the islands' share of the profits has been very small.

When the 1994 Forum opened—its theme "Managing Our Resources"—the three Melanesian states most affected by logging and timber exports had already undertaken measures to control the industry more effectively. An agreement among the three states, Australia, and New Zealand to develop a common code of conduct and surveillance measures governing logging and timber exports was endorsed by the Forum.

* * *

From the Forum perspective, significant progress has been made on many issues. Nuclear testing has ended in the Pacific. The removal of nuclear weapons from U.S. surface ships has reduced the political complications involved in accepting U.S. Navy port calls. The Treaty of Rarotonga will soon have been accepted by all five nuclear powers. Waste disposal in the Pacific has come under greater international regulation. Distant-water fishers have become considerably more respectful of island sovereignty, while at least some island states have begun to develop tuna fleets and canneries of their own.

To be sure, Forum pressures were not decisive in bringing about these favorable developments, shaped as they were both by the cold war and its end. Nevertheless, the Forum countries can take satisfaction from their contribution and from having been in the vanguard of the now more widespread movement for protecting the atmosphere and conserving global resources.

Notes

1. FBIS-EAS, August 29, 1984.

2. *Honolulu Advertiser*, September 3, 1986.

3. FBIS-EAS, September 2, 1988.

4. East-West Center, *The Summit of the United States and the Pacific Island Nations* (Honolulu: East-West Center, 1990), p. 13.

5. Article 36, Communiqué of the Twenty-Fifth South Pacific Forum.

6. FBIS-EAS, August 17, 1995.

7. Ibid., November 7, 1980.

8. In the Nauru Agreement, signed by the Federated States of Micronesia, Kiribati, Marshall Islands, Palau, Papua New Guinea, and the Solomon Islands.

10

Conclusion

The Western intrusion into the Pacific island world inaugurated two centuries of change, a process less rapid than the change that has transformed the West, but one that nevertheless has altered island life in many significant ways. However, the traditional order has not been wholly destroyed. The ties of the extended family and the village are weaker, as is the authority of chiefs and Big Men. But they continue to exist, and land ownership remains dominated by traditional patterns, the inroads of commercialization and modernization notwithstanding. The churches still play the central part in the social order that they have long since lost in the West. Traditional culture is vigorously expressed in art, handicrafts, music, and dance; if some of the stimulus to preservation has been provided by tourism, this does not necessarily detract from the results.

The end of World War II ushered in a period of marked political change in the South Pacific, bringing independence to 14 island states and considerable autonomy to the remaining territories but leaving them all economic dependents of former rulers. The generally gradual, peaceful, and consensual transition to this new status eased the adjustment to the political requirements of independence and fostered cooperative relations with old metropoles. In the wider world, however, the island states were notable only in their remoteness, small size, and limited resources.

With the support of Australia and New Zealand, the island states have made the most of their inevitably limited claim on world attention. The Forum has been a major asset, fostering amity and cooperation among its members, providing them with technical services, and, through its growth and cohesion, becoming a voice of some consequence on global atmospheric and resource issues. The cold war helped also; perhaps uniquely in the third world it was an almost unmixed blessing. Troubled neither by surrogate wars nor by ideological battles in the domestic forum, island governments were not slow to take advantage of opportunitiesto manipulate cold war competition.

The bargaining power of the island states has been much reduced by the end of the cold war which, however, has not reduced the global competition for aid from the advanced countries, many of them confronting acute economic difficulties of their own. Reflecting on the implications of these developments, Australian minister for Pacific Island affairs Gordon Bilney has asked, "Does it not mean that it is up to us; that, paradoxically in this increasingly interdependent world, we are much more on our own?"[1]

More on their own, but not entirely so, the Pacific island polities still have assets and friends. Their waters not only provide a large and growing proportion of the commercially important tuna catch, they also constitute one of the world's few fishing grounds where the catch can be significantly increased without threatening the survivability of the resource. In time, as technology improves, sea bed minerals, not commercially exploitable in 1996, may become another asset. In international forums, as long as they remain cohesive, the island polities make up in numbers something of what they lack in size.

Their membership in the Forum and their concern with their Pacific identity assure the continued commitment of Australia and New Zealand to the region and their support of island development efforts. Although the United States has already terminated its two AID missions in the region and is phasing out its bilateral programs, it still provides some assistance through multilateral programs concerned with such matters as education and health. Its relations with the freely associated states and its flag territories ensure a continued U.S. presence in the region. France, seemingly as dedicated as ever to its imperial image, is much more active in the region, seeking closer relations with the independent island states both for itself and for its territories. Japanese involvement in the South Pacific continues to increase, reflecting both Japan's commitment to a role in its region commensurate with its great-power status and its position as an important consumer of South Pacific tuna and forest products. Since 1987 when then–foreign minister Kurinari Tadashi pledged Japan's support for island political stability and well-being (the so-called Kurinari Doctrine), Japanese aid to the region has increased substantially. The members of ASEAN are also looking with greater interest toward their small southern neighbors. Mindful of its common border with Papua New Guinea, Indonesia has underlined its goodwill by sponsoring PNG's observer status in ASEAN, and, more recently, its membership in APEC. Beijing and Taipei (with which four of the island states maintain diplomatic relations) continue to compete in the region as elsewhere for diplomatic and economic ties.

Even assuming a continued flow of external support, island governments are confronted with a host of difficulties, some likely to get worse. Corruption is a serious problem. Not just in Fiji and New Caledonia but elsewhere as well, cleavages between population groups can be a source of strain or worse, persisting in Papua New Guinea, for example, in both tribal warfare and secessionism. Urbanization has frequently outstripped the services, facilities, and employment opportunities, for youth in particular, that urban areas might be expected to provide. Slow growth and population pressure threaten government ability to continue providing relatively high levels of health care, sanitation, and education. Policymakers have yet to discover the path to lean and efficient government, fiscal stability, and self-sustaining growth.

Despite these and other differences, the blend of Western parliamentary institutions and traditional practice inaugurated during the decolonization process has largely stood the test of time. Measured against other parts of the developing world, or even the world as a whole, the Pacific islands—no Paradise now or ever—have been notably free of blood baths and extremes of repression. A record of adherence to constitutional norms, peaceful acceptance of the results of free elections, and civilian authority has been broken only by Fiji, and even there, civilian constitutional government has been restored.

Note

1. In an address to the Foreign Correspondents Association, Sydney, June 15, 1994.

Appendixes

Appendix 1: Vital Statistics

	Land area[a] (km2)	EEZ[a] (000 km2)	Population[b]
American Samoa	197	390	55,223
Cook Islands	240	1,830	19,124
Federated States of Micronesia	701	2,978	120,347
Fiji	18,272	1,290	764,382
French Polynesia	3,265	5,030	215,129
Guam	541	218	149,620
Kiribati	690	3,550	77,853
Marshall Islands	179	2,131	54,031
Nauru	21	320	10,019
New Caledonia	19,103	1,740	181,390
Niue	259	390	1,906
Northern Marianas	471	777	49,799
Palau	494	629	16,366
Papua New Guinea	462,840	3,120	4,196,806
Solomon Islands	28,369	1,340	385,811
Tokelau	10	290	1,523
Tonga	699	700	104,788
Tuvalu	26	900	9,831
Vanuatu	11,880	680	169,776
Wallis and Futuna	255	300	14,338
Western Samoa	2,935	120	204,447

a. Te'o I.J. Fairbairn et al., *The Pacific Islands* (East-West Center, Honolulu, 1991), p. 6.

b. Central Intelligence Agency. *The World Fact Book, 1994* (Washington, D.C., 1994), passim.

Appendix 2: Social Indicators of the Independent Island States

	Life expectancy at birth	Adult literacy rate	Mean years of schooling	Percent of population with access to health services	Percent of population with access to to safe water
Cook Islands	69.8	99	8.4	100	99
Federated States of Micronesia	64.1	81	7.6	75–80	30
Fiji	63.1	87	6.8	98	92
Kiribati	60.2	93	6.1	85	65
Marshall Islands	61.1	91	8.5	95	50
Nauru	55.5	90	7.3	100	90
Niue	66.0	99	8.3	100	100
Palau	67.0	98	9.6	75	88
Papua New Guinea	49.6	52	2.1	88	23
Solomon Islands	60.7	23	28	80	61
Tokelau	68.0	99	n/a	100	100
Tonga	69.0/	99	7.1	100	100
Tuvalu	67.2	98	6.8	100	100
Vanuatu	62.8	64	4.0	n/a	87

Source: United Nations Development Program. *Pacific Human Development Report* (Suva, Fiji, 1994), pp. 74–75.

About the Book and Author

This accessible volume provides a brief introduction to the institutions, policy concerns, and international roles of the Pacific Islands. Evelyn Colbert expertly paints an overall picture of the region using broad brush strokes, complementing the mostly specialized literature available about the South Pacific.

The Pacific Islands traces the islands' political transition from Western colonies to the mostly independent polities of today. The book begins by describing the geographic and cultural characteristics that distinguish the major island groups of Polynesia, Melanesia, and Micronesia. An engaging history of Western perceptions about the islands follows, starting with the eighteenth-century romantic vision of the South Pacific as Paradise, an idea still prevalent today. The heart of the book examines issues of governance, encountered first by colonial administrators and then by the islanders themselves. Colbert concludes her study with a discussion of international issues faced by the islands, such as French nuclear testing on the Muroroa Atoll and the destructive exploitation of the islands' natural resources.

Evelyn Colbert is a former Deputy Assistant Secretary of State for East Asian and Pacific Affairs. She has taught at the State Department's Foreign Service Institute and the Nitze School of Advanced International Studies at Johns Hopkins University.

Index

Milton Keynes UK
Ingram Content Group UK Ltd.
UKHW022109141024
449569UK00031B/1840

9 780813 332864